Creating With Concrete

Opening and Leader Papers of the
Proceedings of the International Congress
held at the University of Dundee, Scotland, UK
on 6-10 September 1999

Edited by

Ravindra K. Dhir
Director, Concrete Technology Unit
University of Dundee

Peter C. Hewlett
Director, British Board of Agrément

and

M. Roderick Jones
Senior Lecturer, Concrete Technology Unit
University of Dundee

Thomas Telford

Published by Thomas Telford Publishing, Thomas Telford Limited, 1 Heron Quay, London E14 4JD.

URL: http://www.t-telford.co.uk

Distributors for Thomas Telford books are
USA: ASCE Press, 1801 Alexander Bell Drive, Reston, VA 20191-4400, USA
Japan: Maruzen Co. Ltd, Book Department, 3–10 Nihonbashi 2-chome, Chuo-ku, Tokyo 103
Australia: DA Books and Journals, 648 Whitehorse Road, Mitcham 3132, Victoria

First published 1999

The full list of titles from the 1999 International Congress 'Creating with Concrete' and available from Thomas Telford is as follows

- *Creating with concrete*
- *Radical design and concrete practices*
- *Role of interfaces in concrete*
- *Controlling concrete degradation*
- *Extending performance of concrete structures*
- *Exploiting wastes in concrete*
- *Modern concrete materials: binders, additions and admixtures*
- *Utilizing ready-mixed concrete and mortar*
- *Innovation in concrete structures: design and construction*
- *Specialist techniques and materials in concrete construction*
- *Concrete durability and repair technology*

A catalogue record for this book is available from the British Library

ISBN: 0 7277 2833 4

Printed and bound in Great Britain by MPG Books, Bodmin, Cornwall

PREFACE

Concrete is the key material for Mankind to create the built environment, the requirements for which are both demanding in terms of technical performance and economy and yet greatly varied from architectural masterpieces to the simplest of utilities. This presents the greatest challenge and the question is how best to advance concrete and create imaginatively.

In response, the Concrete Technology Unit (CTU) of the University of Dundee organised this Congress following on from its established series of events, namely, Concrete in the Service of Mankind in 1996, Concrete 2000: Economic and Durable Concrete Construction Through Excellence in 1993 and Protection of Concrete in 1990.

Under the theme of Creating with Concrete, the Congress consisted of five Seminars: (i) Radical Design and Concrete Practices, (ii) Role of Interfaces in Concrete, (iii) Controlling Concrete Degradation, (iv) Extending Performance of Concrete Structures and (v) Exploiting Wastes in Concrete, and five Conferences: (i) Modern Concrete Materials: Binders, Additions and Admixtures, (ii) Utilising Ready-Mixed Concrete and Mortar, (iii) Innovation in Concrete Structures: Design and Construction, (iv) Specialist Techniques and Materials for Concrete and Construction and (v) Concrete Durability and Repair Technology. In all, a total of 421 papers were presented from 67 countries.

The Opening Addresses were given by Mr Henry McLeish, MP, MSP, Minister for Enterprise and Lifelong Learning, Scotland, Dr Ian Graham-Bryce, Principal and Vice Chancellor of Dundee University, Mrs Helen Wright, Lord Provost, City of Dundee and Professor Peter Hewlett, President of the Concrete Society. This was followed by four Opening Papers by leading international experts; Dr Bryant Mather, US Army Corps of Engineers, Professor Charles F Hendriks, Delft University of Technology, Netherlands, Dr Bjørn Jensen, Danish Technological Institute, Dr Oliver Kornadt, Philipp Holzmann AG, Germany, Professor Jurek Tolloczko, Concrete Society, UK, Mr Michael Téménidès, CIMBÉTON, France and Professor Yves Malier, Ecole Normale Superieure de Cachan, France. The Closing Address was given by Professor John Morris, University of the Witwatersrand, South Africa.

This special volume of the Proceedings contains four opening papers for the Congress and eleven leader papers given at the start of each of the Seminars and Conferences. Collectively, they represent the state of the art in concrete technology and practice and form an outstanding point of reference for practitioners and researchers alike.

The support of 20 International Professional Institutions and 31 sponsors was a major contribution to the success of the Congress. An extensive Trade Fair, participated in by 50 organisations, formed an integral part of the Congress. The work of the Congress was an immense undertaking and all of those involved are gratefully acknowledged, in particular, the members of the Organising Committee for managing the event from start to finish; members of the International Advisory and National Technical Committees for advising on the selection and reviewing of papers; the Authors and the Chairmen of Technical Sessions for their invaluable contributions to the proceedings.

All of the proceedings have been prepared directly from the camera-ready manuscripts submitted by the authors and editing has been restricted to minor changes where it was considered absolutely necessary.

Dundee
September 1999

Ravindra K Dhir
Chairman, Congress Organising Committee

ORGANISING COMMITTEE
Concrete Technology Unit

Professor R K Dhir, OBE (Chairman)

Dr M R Jones (Secretary)

Mr M D Newlands (Joint Secretary)

Professor P C Hewlett
British Board of Agrément

Dr N A Henderson
Mott MacDonald Ltd

Professor V K Rigopoulou
National Technical University of Athens, Greece

Dr S Y N Chan
Hong Kong Polytechnic University

Dr N Y Ho
L & M Structural Systems, Singapore

Dr M J McCarthy

Dr M C Limbachiya

Dr T D Dyer

Dr K A Paine

Dr T G Jappy

Mr P A J Tittle

Mr J C Knights

Mr S R Scott (Unit Assistant)

Miss A M Duncan (Unit Secretary)

INTERNATIONAL ADVISORY COMMITTEE

INTERNATIONAL ADVISORY COMMITTEE (CONTINUED)

NATIONAL TECHNICAL COMMITTEE

Mr P Barber
Manager of the Scheme, The Quality Scheme for Ready Mixed Concrete

Professor A W Beeby
Professor of Structural Design, University of Leeds

Mr B V Brown
Divisional Technical Executive, Readymix (UK) Ltd.

Dr T W Broyd
Technology Development Director, W S Atkins Ltd.

Professor J H Bungey
Professor of Civil Engineering, University of Liverpool

Dr P S Chana
Director, CRIC, Imperial College of Science, Technology & Medicine

Professor J L Clarke
Principal Engineer, The Concrete Society

Dr P C Das
Group Manager, Structures Management, Highways Agency

Dr S B Desai, OBE
Principal Civil Engineer, Department of the Environment, Transport and the Regions

Professor R K Dhir, OBE (Chairman)
Director, Concrete Technology Unit, University of Dundee

Mr C R Ecob
Director Special Services Division, Mott MacDonald Ltd.

Professor F P Glasser
University of Aberdeen

Professor T A Harrison
Technical Consultant, Quarry Products Association

Professor P C Hewlett
Director, British Board of Agrément

Professor J Innes
Director of Roads, Scottish Office

NATIONAL TECHNICAL COMMITTEE (CONTINUED)

Mr K A L Johnson
Director, AMEC Civil Engineering Ltd.

Dr M R Jones
Senior Lecturer, Concrete Technology Unit, University of Dundee

Mr P Livesey
National Technical Services Manager, Castle Cement Ltd.

Professor A E Long
Director of School, Queens University of Belfast

Professor P S Mangat
Head of Research, Sheffield Hallam University

Mr G Masterton
Director, Babtie Group Ltd.

Professor G C Mays
Director of Civil Engineering, Cranfield University

Mr L H McCurrich
Technology Development Consultant, Fosroc Construction

Professor R S Narayanan
Partner, SB Tietz & Partners Consulting Engineers

Dr P J Nixon
Head, Centre for Concrete Construction, Building Research Establishment Ltd.

Dr W F Price
Senior Associate, Messrs Sandberg

Professor G Somerville, OBE
Director of Engineering, British Cement Association

Professor D C Spooner
Director, Materials and Standards, British Cement Association

Dr H P J Taylor
Director, Tarmac Precast Concrete Ltd.

Mr M Walker
Technical Manager, The Concrete Society

Dr R J Woodward
Senior Project Manager, Transport Research Laboratory

SUPPORTING INSTITUTIONS

American Concrete Institute, USA

American Society of Civil Engineers, USA

Australian Concrete Institute

Concrete Association of Finland

Concrete Society of Southern Africa

Concrete Society, UK

Danish Concrete Society, Denmark

Fédération de l'Industrie du Beton, France

German Concrete Association (DBV)

Hong Kong Institution of Engineers

Indian Concrete Institute

Institute of Concrete Technology, UK

Institution of Civil Engineers, UK

Instituto Brasileiro Do Concreto, Brazil

Japan Concrete Institute

Netherlands Concrete Society

New Zealand Concrete Society

Norwegian Concrete Association

Singapore Concrete Institute

Spanish Association for Structural Concrete

Swedish Concrete Association

SPONSORING ORGANISATIONS WITH EXHIBITION

AMEC Civil Engineering Ltd.

Babtie Group Ltd.

Bardon Aggregates

Blue Circle Cement

Blyth & Blyth

British Board of Agrément

British Cement Association

Building Research Establishment

Castle Cement Ltd.

Cementitious Slag Makers Association

CIMBÉTON, France

Du Pont de Nemours International S.A., Switzerland

ECC International Ltd.

Elkem Ltd. (Materials)

Fosroc International Ltd.

Grace Construction Products

HERACLES General Cement Co., Greece

John Doyle Group

Lafarge Aluminates

L M Scofield Europe Ltd.

Minelco Ltd.

Mott MacDonald Ltd.

O'Rourke Group

Ove Arup and Partners

SPONSORING ORGANISATIONS WITH EXHIBITION (CONTINUED)

Readymix (UK) Ltd.

Rugby Cement

Scottish Enterprise Tayside

Sika Ltd.

SKW - MBT Construction Chemicals

Thomas Telford Publishing Ltd.

United Kingdom Quality Ash Association

W A Fairhurst & Partners

ADDITIONAL EXHIBITORS

Christison Scientific Equipment Ltd.

CMS Pozament Limited

The Concrete Society

David Ball Group plc.

E & FN Spon

Flexcrete Ltd.

Germann Instruments A/S, Denmark

Natural Cement Distribution Limited

Palladian Publications Ltd.

Quality Scheme for Ready Mixed Concrete

UK Certification Authority for Reinforcing Steel

Wacker-Chemie GmbH, Germany

Wexham Developments

CONTENTS

OPENING PAPERS

REALIZING THE POTENTIAL OF CONCRETE AS A CONSTRUCTION MATERIAL

B Mather

US Army Engineer Waterways Experiment Station

United States of America

ABSTRACT. Concrete is international, but made locally; has infinite variability, but can be made very uniform; and can be made to last as long as you want it to. Therefore, what is needed to more fully realize its potential as a construction material is to understand what we want it to do, learn how to make it so it will do so, use available methods to restrict undesired variability, consider the ethical and environmental aspects of its use, and help the people who are making it to do it better.

Keywords: Concrete, Durability, Environment, Ethics, Service life, Specification overkill, Statistical methods, Uniformity, Variability

Dr Bryant Mather is Director of the Structures Laboratory, U.S. Army Engineer Waterways Experiment Station, 3909 Halls Ferry Road, Vicksburg, MS 39180-6199, USA. He is a Past-President of ACI and ASTM and has served on numerous ACI and ASTM committees. He is a prolific author and editor, and has received numerous awards and honors from ACI, ASTM, and other institutions.

INTRODUCTION

When the question of "realizing the potential of concrete as a construction material" was suggested to me as the subject of these remarks, my first reaction was, what is there that has been built using construction materials that hasn't been built using concrete? What unrealized potential does concrete have as a construction material? Or perhaps one first needs to define "construction" — and then "construction material." Stein (1993) includes in "Construction Categories" a list of building types and uses ranging from "Animal Clinic," "Bank," "Dog Kennel," "Factory," "Garage," "Jail," "Museum," "Roundhouse," "Stable," and "Theatre," to "Warehouse." I believe that examples of all of these have been built using concrete.

I therefore concluded that perhaps the opportunity presented by this title is not so much to talk about how to construct some category of construction that has never before been produced using concrete, but rather describing how to do concrete work of the kind that the human race has been engaged in to some degree or other for the past eight millennia (Malinowski and Friefelt 1993) but doing it better so as to serve better, at lower social cost. We need, inter alia, to learn how to better realize the potential of concrete as a construction material by producing it with the use of less energy and smaller emissions of greenhouse gases.

Even as was the case with these remarks, in 1987 the organizers of the First International Conference on Concrete Durability invited Katharine Mather and me to speak at the opening session. On that occasion, the version that was published had the title "Reflections on Concrete Durability and on International Conferences on Concrete Durability" (Mather and Mather 1987). In it, we reiterated a number of well known relevant facts about concrete. I restate them in slightly revised form as an introduction.

Concrete is international and as we have air to breathe, water to drink, earth to grow plants in; it provides for all that which the 1976 National Ready Mixed Concrete Association's U.S. Bicentennial bumper sticker said: "Concrete—the Foundation of America," and that can now be expanded to "Concrete — the Foundation of Civilization."

Concrete is made locally. The mixing water, with negligible exceptions (perhaps when concrete is made on the moon); the aggregates usually; the cement often, are or are made from one or more locally available natural raw materials, after various degrees of processing.

Concrete has infinite variability. The varieties of concrete that are intentionally produced are more numerous than the known varieties of natural sedimentary rocks. Concretes have been — and will be — required to be produced having densities from as low as 160 kg/m^3 (10 lb mass per ft^3) to as high as 4800 kg/m^3 (300 lb mass per ft^3); that have unconfined compressive strengths from as low as 0.35 MPa (50 psi) to as high as 275 MPa (40,000 psi); that are of any color — white, black, pink, turquoise, chartreuse, or as one Corps of Engineers officer demanded, "sand tan"; we await the demand for "olive drab."

Concrete can be made to be very uniform. ACI Committee 214 was organized in 1946. It told concrete makers and users how to use strength tests to measure uniformity of concrete as produced and still does. In 1979, ASTM Standard Method C 917 for Evaluation of Cement Strength Uniformity was adopted. For many years ASTM Specification C 33 has put rather restrictive limits on uniformity of grading of fine aggregate in terms of fineness modulus. There are uniformity requirements on pozzolans given in ASTM Specification C 618.

Concrete can be made to last as long as you want it to. Even as some relatively fragile sedimentary rocks exist that have survived some hundreds of millions of years, so there are concretes that, if they remain in relatively mild environments, could last forever or at least as close to forever as we can measure. The duration of the life of a concrete with its relevant properties not sufficiently degraded so as to impair or jeopardize its ability to serve its intended purpose, is not usually a function of the absolute level of any universally relevant property of the concrete such as slump, density, strength, modulus of elasticity, or color but is almost always a function of the interaction of some property or properties, the required levels of which depend on the severity of the environment to which the concrete will be exposed in service. One need not worry much about resistance to freezing and thawing in the tropics, except in facilities using refrigeration; or at the poles, except in facilities using intermittent heating. One need not worry about acid attack, abrasion attack, sulfate attack, attack by boring or burrowing molluscs, or any of the myriad hazards described by Kleinlogel (1950), Biczok (1976), or others except where they exist and impact the concrete of interest. And where these potentially destructive influences can be and are brought to bear on the concrete, if their action and its intensity and duration are properly anticipated, the concrete can be rendered adequately resistant so as to merit the designation Durable, even when as demanded of us by authorities dealing with the storage of nuclear waste, means not significantly degraded for many centuries or in some cases hundreds of centuries.

With these considerations in mind, what is there to say about the unrealized potential of concrete as a construction material? In my view, several things: First, what do we need to do that we are not doing that, if we did it, would greatly reduce the number of cases of premature degradation of concrete in service. Too much concrete that should last at least 50 or 100 years with minimal damage is showing serious damage in a tenth or less of that time. And, next, what do we need to do that we are not doing that, if we did it, would greatly reduce the cost of concrete by not paying extra to provide resistance to aggressive influences that have negligible likelihood of being encountered by the concrete in its intended service life. We now, often, pay a lot to provide frost-resistant concrete for use in regions where natural freezing has not taken place for millions of years. If we only ordered what we need and can benefit from, we'd have fewer things to look out for through quality control and quality assurance, and that, too, would contribute to lowering the cost.

One important matter, therefore, is to know how properly to assess the nature and severity of the environment of service; to get a reasonable estimate of intended service life; to get a reasonable assessment of how much degradation is tolerable structurally and, when relevant, aesthetically. One must then, from these and known relationships of environmental influences and concrete composition and constitution, have and use procedures and criteria by which to select materials, proportions, and construction practices that will, with reasonable probability of success and compliance with the owner's budget for the work, give concrete that will not prematurely deteriorate.

Most of the knowledge to do this exists, some in relatively quite refined states. But the manner in which this knowledge is used in an attempt to make it an effective tool for makers and users of concrete is primitive. Today we talk about reactive vs non-reactive aggregate and high-alkali vs not high-alkali cement as if all structures, in all environments, involving all ambient temperatures and temperature ranges, in all degrees of moistness or dryness needed the same level of protection. There is no relaxation of the specified limit of 0.60 percent equivalent Na_2O maximum on cement as a function of environmental severity, degree of reactivity of aggregate, or anticipated level of permeability of concrete. An aggregate is either reactive or not; if not, nothing need be done, no precautions taken; if reactive, then the precautions must be taken without reference to degree of reactivity or degree of severity of exposure.

Today we talk about need for resistance to freezing and thawing. If the concrete will ever likely freeze when any of the spaces in it that can hold freezable water are more than 91 percent water filled, we need air-entrainment, sound aggregate, and protection against freezing until moderate maturity is achieved. And, air-entrainment means an air content that will give an air-void system with a bubble-spacing factor, $\bar{L}$, less than 0.2 mm (0.008 in.) regardless of whether we have one or one hundred cycles of freezing and thawing per year and regardless of whether the rate of freezing is fast or slow or whether the minimum temperature achieved 10 mm below the exposed concrete surface is minus 5 °C or minus 50 °C.

In 1987 (Mather and Mather), we decided to refer to these sorts of approaches as specification overkill. It involves, much of the time, paying more than needs to be paid to buy a level of protection that, in many cases, cannot ever be needed and from which no good can ever be derived. Ever in the preceding sentence means not more often than once in a million years. Today we have computers that can be run by most bright junior-high-school scientists that could be used to take the knowledge that does exist and derive from it reasonable criteria for matching degree of precautions needed or degree of restriction required of the contractor in his exercise of choice among materials, mixture proportions, and practices, so that only limited precautions and light restrictions are called for when the nature and intensity of the anticipated environmental hazards and insults the concrete must resist are slight or of short duration in time. We can then reserve the levels of precautions we now take for all levels of severity for those cases where the hazard is really very severe and protection is needed for a very long time. We might well, by so limiting these requirements in their applicability, be able to increase their degree of protection so as further to minimize the probability that they will ever be less effective than needed. My challenge therefore is: Since concrete need not deteriorate, ensure that it doesn't, but do so without raising the cost more than necessary; stamp out specification overkill.

VARIABILITY

Another area in which there is significant room for improvement in concrete technology in practice in the degree to which we realize the potential of concrete as a construction material is the application of the theory of probability to the understanding of, specification of, and interpretation of test data on the measured properties of concrete. The application of the theory of probability to the ready-mixed concrete industry was described by Stanton Walker (1944, 1944a) in a section of the publication *Control of Quality of Ready-Mixed Concrete* entitled "Over-Design to Meet Specifications." Tables were given showing the average strength required for operations with different coefficients of variation so that there will be 9 chances out of 10 or 99 chances out of 100 that no test will fall below the specified strength. He remarked that these relations "demonstrate forcibly that investment in good control is self-liquidating." In 1946, the ACI Advisory Committee, under his chairmanship, recommended the formation of a new technical committee entitled, "Evaluation of Results of Compression Tests of Field Concrete." The mission of the committee was stated as follows: The assignment is to analyze data of compressive tests of samples of concrete taken in the field to determine the nature and extent of variations in them for different job conditions and to develop measures of the reliability of such tests as bases for the design of concrete proportions. (Taken from a letter from Stanton Walker to Harvey Whipple, November 27, 1946.)

The committee produced a standard which as of 1965 was called "ACI Standard Recommended Practice for Evaluation of Compression Test Results of Field Concrete" (ACI 214-65), which had as its abstract:

> "Statistical methods provide valuable tools for assessing results of strength tests, and such information is also of value in refining design criteria and specifications. The report discusses briefly the numerous variations that occur in the strength of concrete and presents statistical methods which are useful in interpreting these variations. Criteria are offered that can be used to establish specifications and maintain required uniformity. An appendix presents a simplified version of statistical quality control procedures."

In 1973 ACI SP 37, "Realism in the Application of ACI Standard 214-65," was published. The preface to this SP included the following:

Standard ACI 214, since its first promulgation, has produced something of a revolution in the field of evaluation of concrete strength tests, and, when applied properly, has resulted in considerable improvement in the quality, and especially the uniformity, of the concrete on projects where it was used. However, for reasons that are explored in some of the papers in this symposium, acceptance of the ideas and techniques contained in the Standard has been slow and reluctant in some quarters. This situation has been the subject of much discussion at meetings of Committee ACI 214 in recent years.

At the meeting in Denver, in March 1971, the conclusion was reached that what was needed was a strong program of education. As a first step in this direction, the decision was made to hold a symposium at the fall meeting. A committee consisting of Howard T. Arni, Federal Highway Administration, Co-chairman; Edward A. Abdun-Nur, Consulting Engineer, Co-chairman; Francis J. Principe, Principe-Dana, Incorporated; and V. M. Malhotra, Canada Department of Energy, Mines, and Resources, was appointed to organize the symposium.

The Symposium was held on November 4, 1971, in Buffalo, New York.

Mr. Abdun-Nur's two papers, "Control of Concrete Mixes," which was co-authored by Joseph J. Waddell and appeared in the March 1959 ACI *Journal*, and "How Good Is Good Enough?" in the January 1962 *Journal*, have become classics in the field of evaluation of strength tests and are still extremely helpful to those who wish to make use of Standard ACI 214. The committee hopes that the inclusion of these two papers and of Standard ACI 214 itself as Appendices along with the other papers will make this volume a very useful guide to those who are involved in problems of evaluation of strength tests of concrete.

Regrettably, much of what was said in 1973 about the application of probability is still true. What was said about what was needed is. . .education is still true — and always will be true. We could all benefit from reading Ed Abdun-Nur's paper on "How Good Is Good Enough?" again (Abdun-Nur 1962).

In 1993 I was invited to write a Foreword to a book called *Concrete Mix Design, Quality Control and Specification* by Ken Day (1995); in it I said:

> "I was once required to attend night courses at the laboratory in which the Director, Mr. Charles E. Wuerpel, taught a course in concrete mix design. I failed to learn to do it — too much arithmetic. This was in 1942. This book, I believe, tells the reader that there is a great deal more arithmetic now than then, but don't worry; the computer will take care of getting it done, provided you instruct the computer properly.
>
> "Somewhat later I became convinced that most of the time one could make concrete of appropriate quality for a given use with locally available materials. I once upset some nuclear physicists by suggesting that 'sidewalk' concrete from the local neighbourhood ready-mix plant, if uniform, would make just as good a biological shield around a nuclear reactor as concrete with high-density aggregate imported from a thousand miles away. I suggested that the extra volume needed would cost less than the shipping cost of a smaller volume of expensive aggregates. Ken Day understands this (section 1.2).
>
> "It has been one of my theses that properly proportioned concrete cannot be over vibrated — but one needs to proportion concrete for 'excessive' vibration if one wants to vibrate it excessively. Ken knows this too (section 1.3). I have also argued that one can design structures (including pavements) for whatever loading is relevant and then select a design concrete strength for that loading (compressive, flexural, or whatever) and proportion a mixture to achieve it; but having done so, it is dumb not to use compressive strength for routine control testing; Ken says this (section 1.7). He knows that hot weather does not directly increase water demand, he just implies it like everybody else. However, unlike everybody else he has a proposal ('equivalent slump,' section 12.1) to get to the truth of the matter, as I have been urging for many years.
>
> "The concept of distinguishing acceptable non-conforming concrete from unacceptable non-conforming concrete that Ken calls 'structurally defective' and 'contractually defective' needs more emphasis; and this book provides it.
>
> "It is good to be reminded of the fact that when worrying about the 'actual' strength of concrete in a structure it is quite proper to worry about its present strength under some conditions and its eventual strength under others (section 5.15). It is also good to read in section 6.1: 'Fortunately the pressure to specify minimum cement contents was resisted.' We had the battle of New Orleans about this in the ACI and a rule was made that if there were to be a minimum cement content there must also be an alternative performance requirement that if met would obviate the need for the prescriptive one.
>
> "I was very happy to see 'Waterproofers' put into quotation marks, since the term implies doing something that is not possible.

> "For these and other reasons I am pleased to commend this book to its readers, in spite of my belief that, in concrete technology, mix should only be used as a verb and design should only refer to selection of properties of concrete structural elements and hence the title should have been 'Concrete Mixture Proportioning.'"

I am told that a new edition of this may appear in 1999.

SOCIAL/ETHICAL ISSUES

In April 1995 the Netherlands Cementindustrie in its magazine "Cement" included the following: "Does the future of concrete have anything to do with ethics? Any human activity has an ethical dimension. This is true of science and technology, as therefore for civil engineering too. The future of concrete will not be affected so much by ethical considerations, but mainly by technical, economic, social, and ecological developments. Nevertheless ethical guidelines can contribute to the attitudes adopted. If the construction industry acts responsibly, with a keen realization of the contribution it can make, concrete as a construction material will have an important role in the future. At the beginning of any technical venture all those involved should ask themselves in principle: 'Will this do more harm than good?' In the many cases where those involved do not consider it incumbent upon them to stand on the sidelines, they should follow some ethical guidelines, which apply to all areas of science and technology."

Among the questions that are properly asked about concrete as a construction material, in addition to the questions about whether what it is to be used to build will do more harm than good — e.g., ICBMs, nuclear power plants, gambling casinos, etc. — is the question of whether concrete or some alternative construction material does more harm to the environment in its production and use. In the United States, there has been created an Environmental Council of Concrete Organizations (ECCO) with its headquarters at the Portland Cement Association. It has produced a folder called "The Surprising Environmental Benefits of Cement" which points out that unlike building materials that rely on scarce or non-renewable resources, concrete is made from three abundant, readily available ingredients: (1) water, (2) aggregate (stone, sand, and gravel, and (3) portland cement (a fine gray powder made from abundant natural materials). A growing list of recycled materials supplements these three basic ingredients. In addition to stone, sand, and gravel, aggregate may contain some recycled materials, such as slag, a by-product of steel-making. And part of the portland cement can be replaced by other materials, such as fly ash, a by-product of coal-burning power plants.

Even the actual process of making cement uses recycled materials. High-energy wastes such as old tyres can be safely recycled as fuel for the cement-making process. Each year, a single cement kiln can recycle one million tyres, conserving fossil fuels and placing less demand on landfill space. It then points out that when you consider that concrete is strong, adaptable, and resistant to fire, water, and weather, there's little wonder it has a longer service life than wood, steel, asphalt, and other construction products. Concrete's durability conserves resources by reducing maintenance and the need for reconstruction. And at the end of its initial service life, concrete can be crushed and reused as aggregate for new concrete, continuing the cycle of environmental benefits. The folder also notes that the thermal mass of concrete buildings saves energy year-round by reducing temperature swings.

During the cooling season, the building will require cooling mainly at night — an off-peak time when electric companies can produce power more efficiently and conserve fossils fuels. During the heating season, a concrete building will cost less to heat than a comparable structure made of different materials. And these benefits last as long as the building itself — which in the case of concrete structures is a very long time.

It is concluded that concrete is one of the single most environmentally friendly construction products currently available.

In another connection, I was once asked to speak to the question of whether or not I was aware of any structure built of concrete by the Corps of Engineers that would have been less of an insult to its environment had it been built of a different construction material. After deliberation and consultation, I concluded that no Corps of Engineers concrete structure I could think of would have been more environmentally friendly if made of some other construction material. When I made this point in a talk, a member of the audience told me of a concrete repair to the tailrace from a penstock at a hydroelectric power dam where when the repair was completed it was noted that fish tended to assemble close to the new concrete surface. I have not been given any explanation of this, but I regard it as tending to validate my conclusion. I suggest that considering ethical, social, environmental ("green") issues will, to an increasing degree, help in the realization of the potential of concrete as a construction material.

PROBLEMS

In 1997 the National Ready Mixed Concrete Association surveyed its membership on technical and production-related problems. Highly rated problems were several quality assurance issues, including improperly conducted fresh and hardened concrete testing and obtaining reports from commercial laboratories. Problems relating to materials listed aggregates and cements as being variable and a lack of communication by the material suppliers when the characteristics of these materials changed. Other highly rated problems related to controlling air content and slump, primarily due to changing compatibility of cements and admixtures. On specification issues, a significant problem was dealing with unreasonable requirements, such as inconsistent prescription and performance requirements.

When asked to describe a recent technical or production problem, about 40% indicated that they were dealing with variations in materials and mixture performance, such as setting time, slump, strength, and air content. Another 30% were dealing with customer complaints regarding concrete flatwork and finishing problems, often related to improper concreting practices. About 20% of the responses related to complying with unreasonable and conflicting specification requirements, such as specifications for specific brand materials, significantly over-designed mixtures, and other durability-related test qualification.

About 30% of production problems related to improper functioning of plant and truck equipment and control systems. Obtaining and retaining trained plant personnel and truck drivers was cited as one of the primary problems. About 10% of the comments related to compliance with environmental regulations and the problems with handling returned concrete, wash water, and storm water.

When asked about how things could be improved, 40% indicated that there needs to be some means to educate and inform engineers, architects, and other specifiers about following national standards while allowing flexibility for local conditions and materials; standardize the submittal process; move toward performance-based specifications; and promote the use of concrete. About 20% indicated that testing issues should be addressed by improving and informing about standard procedures and the qualification of testing personnel. More educational and training programs for ready-mixed concrete producers, including sales training; dealing with material variability; mix designs for high-performance concrete; and other quality control functions were stressed by about 15% of the responses. Addressing durability issues such as scaling and other flatwork problems; alkali-silica reactivity; and concrete permeability specifications were also mentioned. About 9% also indicated that contractor and customer education should be made available, while another 9% indicated that effort should be expended toward updating guides and standards in ACI, ASTM, and other organizations.

CONCLUSIONS

The potential of concrete as a construction material will be better realized when:

1. We improve our procedures for describing what we want so we don't pay for behavior we don't need and can't use, i.e., we learn to specify intelligently.

2. We employ readily available procedures for describing, understanding, and controlling variability of concrete.

3. We consider the ethical aspects of construction and especially the environmental aspects of concrete production and use.

4. We find out and act to alleviate the real life problems of the people who are out there trying to make good concrete.

REFERENCES

ABDUN-NUR, EDWARD A. 1962. "How Good Is Good Enough?" Journal, American Concrete Institute, *Proceedings*, vol 59, pp 31-46 (reprinted in ACI SP 37, 1973, pp 173-187).

AMERICAN CONCRETE INSTITUTE. 1973. *Realism in the Application of ACI Standard 214-65*, ACI SP 37, American Concrete Institute, Detroit, MI, 215 pp.

AMERICAN CONCRETE INSTITUTE 1998. "Recommended Practice for Evaluation of Strength Test Results of Concrete," ACI 214-77 (Reapproved 1997), ACI *Manual of Concrete Practice*, Part 2, 14 pp.

AMERICAN SOCIETY FOR TESTING AND MATERIALS. 1998. Vol 04.02: C 33: "Standard Specification for Concrete Aggregates," C 33-97, pp 10-16; C 618: "Standard Specification for Coal Fly Ash or Calcined Natural Pozzolan for Use as a Mineral Admixture in Concrete," C 618-98, pp 298-300.

AMERICAN SOCIETY FOR TESTING AND MATERIALS. 1998. Vol 04.01: C 917: "Standard Test Method for Evaluation of Cement Strength Uniformity from a Single Source," C 917-95a, pp 417-424.

BICZOK, IMRE. 1976. *Concrete Corrosion and Concrete Protection.* 8th Ed, Akadèmiai Kiado, 545 pp, Budapest.

KLEINLOGEL, ADOLPH. 1950. *Einflüsse auf Beton and Stahlbeton*, 5th Ed, revised by Kurt Walz and Hugo Vierheller, Verlag von Wilhelm Ernst & Sohn, 339 pp, Berlin.

MALINOWSKI, ROMAN, AND FRIEFELT, KAREN. 1993. "The Ubaid Plaster of Dosariyah (5-4000 Years BC)," Swedish Council for Building Research, *Prehistoric Hydraulic Mortar*, D12, 16 pp, Stockholm.

MATHER, KATHARINE, AND MATHER, BRYANT. 1987. "Reflections on Concrete Durability and on International Conferences on Concrete Durability" in *Concrete Durability*, John M. Scanlon, Editor, ACI SP-100, pp 1-5, 2 vol, 2179 pp, American Concrete Institute, Detroit.

NATIONAL READY MIXED CONCRETE ASSOCIATION. 1997. Technical Information Letter No. 529, Silver Spring, MD, 8 pp.

NETHERLANDS CEMENTINDUSTRIE. 1995. "Does the Future of Concrete Have Anything to do with Ethics?" *Cement*, vol 47, no. 4 (April).

STEIN, J. STEWART. 1993. *Construction Glossary*, John Wiley and Sons, Inc., 1137 pp, New York.

WALKER, STANTON. 1944. "Application of Theory of Probability in Designing for Strength Specifications," Technical Information Letter No. 14, National Ready Mixed Concrete Association, Washington, DC (Reprinted in *Rock Products*, March 1944; in *Concrete*, May 1944).

WALKER, STANTON. 1944a. "Control of Quality of Ready-Mixed Concrete," National Ready Mixed Concrete Association, Washington, DC, 71 pp (2nd Ed, 1945, 75 pp; 3rd Ed, 1953, 76 pp).

GLOBAL USE OF CONCRETE AND ITS IMPACT ON THE ENVIRONMENT

Ch F Hendriks

Delft University of Technology

Netherlands

B L Jensen

Danish Technological Institute

Denmark

ABSTRACT. This contribution gives information on the production and use of cement and concrete for all major applications. The durability of these applications is discussed; also the most recent developed knowledge on improvement of durability is given. Concerning the sustainability the most important aspects for the indoor and outdoor applications are analysed: emissions to air, leaching, radiation, dust depletion of raw materials, waste production, noise and vibration and recycling or reuse. Finally, the protection of the environment by the use of concrete is included in this paper.

Keywords: Cement, Concrete, Durability, Sustainability, Regulation, Emission, Leaching, Recycling, Energy

Professor Charles F Hendriks is professor in Materials Science in the faculty of Civil Engineering and Geosciences at Delft University of Technology, Netherlands and is responsible for research in the field of durable and sustainable construction.

B Lykke Jensen is director for the building division of the Danish Technological Institute, Denmark. The core activities of the division are material technology, cleaner technologies, and product certifications, mainly related to concrete, masonry and wooden structures.

PART 1 - GLOBAL CONSUMPTION OF CONCRETE

Introduction

Since the use of concrete is closely related to the modern civilisation, the increase in use will be expected to relate to the growth rate of how the modern civilisation is spread globally. Concrete is well suited for rapid urbanisation. A large number of new housings can be provided in very short time, and concrete structures, like bridges, tunnels etc. are important parts of new infrastructures.

Figure 1 is showing the world production of cement in 1989 and 1995, The cement production increases with a yearly rate of app. 4.5 %, but with considerably variations between the different parts of the world. In the period the production in Asia has almost doubled, While the production in Japan, USA and Europe has gone slightly down. Since the effect from environmental requirements to cleaner production methods, (in the developed countries, app. 1.5 % of the total emission of CO_2, stems from concrete and cement production [1] the cost of transport, manpower etc. is not specifically given in these figures. The figures do not directly reflect the actual use of concrete. Nevertheless, the figures indicates that the use of concrete is increasing world-wide, with an average growth rate close to 4.5 %, and with a very rapid growth rate in Asia, which relates well with the very rapid growth of civilisation in this part of the world. Obviously, this increased use of concrete, like the civilisation itself, effects the environment.

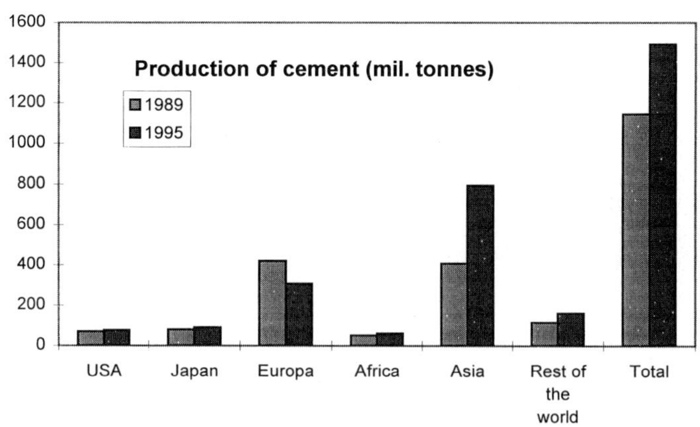

Figure 1 Production of cement 1989 and 1995 (CEMBURO)

Some Environmental Consequences of the Use of Concrete

For 'the environment' the following definition can be adopted: 'The surrounding conditions, including natural- and social conditions that effects human beings, animals and plants over time.'

Bearing in mind, the importance of concrete in the development of modern civilisation, the impact to the environment, originating from the use of concrete is wide and not only limited to the actual concrete structures and the material.

Concrete in housing

The use of concrete facilitates the erections of massive housing concentrations, which in many ways effects the ways of living in the world today. Especially in the developing countries, the use concrete, helps housing of the massive inflow of population from poorer country districts to the cities, which is good, but often these housings are of poor quality and lacks recreation facilities. These psychological and social life conditions in these modern concrete slum-areas do in no way reduce the social problems that the inhabitants may suffer under - on the contrary. Concrete itself cannot be blamed for this - but the way it is used certainly can.

Figure 2 Modern housing area - potential slum area

In many ways it is a shame that concrete has gained this reputation, because if used properly, and in the hands of good architects and town planners, concrete on the other hand is a unique material. It allows creation of unique architecture, and can help creating townships and infrastructure with permanent values for its users.

One could argue that both housings in good as well as in poor quality is possible without the use of concrete, which obviously is correct. But since concrete is relatively inexpensive, has good characteristics for both fire-resistance and sound-damping, requires only very limited maintenance, and is an ecological more sound product than other conventional materials like Steel an glass [2], the development as we sees it today, and its financial, social and ecological consequences, is closely related to the use of concrete.

Figure 3 An example of excellent architecture, Arken Museum for modern art in Copenhagen

Concrete in infrastructure

In the planning of the Great Belt crossing in Denmark, the construction impact on the environment turned out to be one of the most important issues in the political decision process

The Great Belt crossing, which opened in 1998 is nearly 18 km long. It comprises a suspension bridge (East bridge) with the worlds second-largest free span of 1624 m, an 8 km long bored railway tunnel, an artificial island and a 6,6 km long low bridge (West Bridge) for railway/highway. The East Bridge has concrete pylons, columns and anchor-blocks, the Tunnel has high strength concrete linings, and the low bridge is all concrete.

Figure 4 The Great Belt crossing

The Great Belt is the main source to the exchange of fresh salty water from the North Sea and the waters of relatively closed Baltic Sea. The flow conditions in the belt are stratified, meaning that the cold salty waters from the North Sea runs toward the Baltic in the lower layer, while the fresh, yet polluted waters from the Baltic, runs in the opposite direction in the upper layer. This exchange is vital to the continuation of life in the Baltic, which is polluted by the rivers from the surrounding countries. Therefore much effort was made to simulate to what degree the many columns from the low-bridge would cause undesired mixing of the two layers, and thereby disturbing the balance of life in the Baltic. The question turned out to be a major environmental issue, which delayed the political go-ahead, until it was decided to carry out massive compensation dredging to ensure an undisturbed flow-balance in the Belt. [3]

Another environmentally related issue in planning of new infrastructure, is the question whether a new crossing contributes to more or lesser CO_2 emission and consumption of energy in total. The better traffic conditions, and thereby more traffic generates added CO_2 emission and also the construction and fabrication of the crossing generates CO_2 emission. On the other hand, the existing sea- and air-traffic is either reduced or closed down. Based on a life-cycle assessment, it has recently been estimated that a coming crossing between Denmark and Germany over the Baltic will lead to a net-reduction in energy consumption, and thereby CO_2 emissions, despite an increased traffic load.[4]. A similar tendency has been recorded at the recently opened Great Belt crossing.

Concrete in other civil works structures

The building of damns is another example to where the use of concrete has massive influence on the environment. Dams provides substantial savings in fossil energy consumption, but on the other hand changes the lifeconditions for the local inhabitants drastically - for good or for bad. [5]

Finally it can be mentioned that concrete is also well suited for enclosures against acute environmental damages. The enclosure of the damaged nuclear reactor at Tjernobyl in Ukraine, is a good example.

Proper Use of Concrete - More Durable Constructions

Although concrete is considered to be one of the most durable, of our normally used construction materials, concrete has a limited lifetime. Concrete deteriorates, and eventually concrete constructions are both left to ruin, and thereby causing harm to the social environment in cities. It is broken down in and put in deposits, and thereby becomes a major consumer of the increasingly limited deposit sites of the world, or it may be recycled as building material in other constructions, typically in the base layers of roads.

Even though there is a positive trend towards recycling of old concrete; in Denmark practically all demolished concrete is crushed an re-used, the process requires energy. Some of this energy might have been saved, had the structures had a longer lifetime.

Improving the lifetime can be achieved both by improving the constructions technically, but before all also by applying architecture and town planning with more permanent value, as often is seen. This is a general statement, which is not limited to the use of concrete. However since concrete is so well suited for rapid development of both infrastructure and in particular new housings, it especially applicable.

There are many ways to technically increase the durability of concrete structures, and the most important ones are:

- Adopt a recipe that provides durable concrete.
- Use skilled workmanship for casting.
- Proper design

Durable concrete mix

To provide durable concrete mix, the concrete first of all must dense without cracks, is able to protect the reinforcement against chloride-attacks and prevent carbonation. It must offer frost resistance, and it must be workable to provide perfect casting, to mention the most important factors. In Denmark the development in this field, has been boosted by the two major bridge projects, The Great Belt Crossing and the Øresund Link.

The nearly 16 km long Øresund Link, comprises a 1,1 km long cable-stayed bridge, a 6,7 km long approach bridge, an artificial island, and a 3,5 km long submerged tunnel. The pylons and columns for the bridges and the submerged tunnel are concrete structures. Furthermore the Link includes dozens of smaller highway- and railway bridges, several kilometres of cut- and cover tunnels, retaining walls etc.

The design life for both crossings is 100 years, which gave rise to new concrete recipes and procedures for workmanship. The recipes should provide a mix that would be more resident to frost and chloride diffusion, than had been seen with conventional types at the time. The consulting engineers the contractors and the manufacturers however were untrained in the used of the new mixes. The mixes required more careful production-, and casting methods than with conventional mixes to maintain a good and dense structure, which led to some difficulties in the initial stages in the project. In the Øresund Link, the concrete requirements where modified, taking account of the experience gained at the Great Belt. [6]

Proper casting

Proper casting of concrete requires skilled workmanship. The list of damages to a concrete structure caused by poor workmanship and inadequate casting procedures are endless. The most often seen are ineffective compaction causing honeycombs and cavities, early age cracking due to terminal strains and ineffective evaporation protection, and inadequate coverage of reinforcement due to inaccurate placement.

Proper design

The structural design must be such that it facilitates durability. This is a well-known fact, but nevertheless poor design is a major contributor to poor durability of concrete structures. Too often the design of a structure is determined purely by the statics, and little attention is given to apply design that will minimise future needs for maintenance. There are at least two factors that will have to be considered, if the design shall provide a durable structure. Details must be deigned to avoid harmful effects from the environment, which in practice often means to ensure an effective water run-off, and the design must be as easy to cast as possible. There can be given many examples of designs that causes a problematic casting process, where among are the use of surfaces with strong inclination, and the use of dense reinforcement. Often these difficult designs cannot be avoided, but too often it can, had the designer had sufficient training in casting difficulties.

PART 2 - CONCRETE IN RELATION TO ECOLOGY [7, 8]

Introduction

Concrete is manufactured from finite raw materials, including the high energy-consuming component of cement. It is a construction material that meets most durable construction premises. The reasons for this are: Raw material supplies for concrete are far from depleted; Concrete has long life and requires little maintenance; If properly applied, concrete can be upgraded after the first life cycle to a high-grade multiple-use aggregate. Finally, concrete is not only highly suited for use as a construction material, but in case of demountable construction, it can also be fully reused as a building component or even as a building,

The energy content of materials, including concrete, and the reduction of energy consumption in the manufacturing process is considered within the 'durable construction' framework. Hence, it is not only the material's durability and application that count in current concrete manufacture. We already know concrete is highly durable. Yet we are presently dealing with aspects like raw material availability, the release of harmful substances during manufacture, the environmental aspects of transport, reuse potential, and solutions for residue matter and waste. This section looks into the ecological angle of concrete.

Using concrete as a construction material offers certain advantages, yet concrete also has less positive environmental properties. A better environment is not proclaimed by law, but comes about by working at it on a daily basis, particularly within the construction community.

Concrete and its Applications

Concrete is definitely the most commonly used construction material of the century. It is used in residential and public building construction, in the agricultural sector, in road construction, and in hydraulics. Prefabricated products have been much used over the past few years. Annual concrete manufacture in the Netherlands amounts to 15 million m^3 (35 million tons), 50 % of which is prefab.

One of the key questions of this paper is then: How to make sure the use of concrete results in minimum environmental attack?

In most countries, construction and demolition waste (CDW) constitutes a major waste stream. We are talking here about an average of 0.7 tons per capita per annum. By comparison, the household waste quantity amounts to approximately half of that.

Regulations

There are few regulations in existence directly affecting the environmental aspects of the construction material concrete. Regulations do exist with regard to the safety and factory aspects of construction material manufacture. Environmental considerations presently play a major part in granting permits to companies. The Building Materials Regulation, dealing with loads on soil and water caused by construction materials, has recently come into force in the Netherlands and in Flanders. Under the terms of the Building Materials Resolution, leaching of environmentally hazardous substances from construction materials that can contaminate the soil or the groundwater should remain limited or be reduced (principle of marginal loading).

Concrete is manufactured in the same way all over the world. Notably from minerals: sand, gravel, crushed rock, soft limestone (marl). Some countries increasingly impose limitations on mineral extraction because of the disruption in the landscape. Requirements with regard to redesigning the area are also becoming increasingly severe.

After marl has been extracted, the pits are redesigned into a scenic area. The top layer, initially excavated for winning purposes, is dumped in the shape of a hill, thus (re)transforming it into a scenic/recreation area.

Not all-environmental regulations are issued by the government. The business community also plays a key role in putting environmental legislation into practice. Standardisation bodies in various countries have drafted directives on waste aggregates in concrete and on liquid-proof concrete.

Possibilities of reuse

When a concrete structure is no longer functional and cannot be adjusted to a new function, the structure must be removed. Yet in case of demountable construction, reuse can take place at the building component level. If not, the alternative is demolition where reuse takes place at the material level. The Netherlands has some seventy CDW-processing plants, which upgrade stone-like waste (12 Mton per annum) to concrete aggregate, masonry aggregate, or mixed aggregate (a mixture of concrete and masonry aggregates).

The Dutch Government has put a ban on CDW-dumping. Ninety per cent of CDW is currently reused. Sixty-five per cent of the stone-like materials is applied as an aggregate, predominantly as a paving material in road construction.

For each ton of aggregate, 2 m^3 of washing water is used. The water is reused in a closed-loop system. Different types of material are produced during CDW-processing, such as metals, timber, synthetics, and sludge. Most of these materials can be reused. Only a minor part is incinerated or disposed of.

Reusing waste, as an aggregate in concrete is much promoted. A whole range of possibilities has been created owing to new technological developments and laws.

Recycled aggregates

In principle, recycled aggregated can be used as a gravel replacement in concrete. Eighty per cent of CDW-waste consists of stone-like material, predominantly concrete and masonry rubble. By applying selective demolition techniques, a waste stream becomes available from which concrete and masonry aggregates of acceptable quality can be produced. Hence the waste is put to a higher-quality allocation than when reused in embankments or pavings, as frequently happened in the past. Twenty per cent of the gravel in concrete can be replaced by such aggregates without affecting concrete properties. Recycled aggregates are used on a small scale since aggregate of adequately controlled quality is lacking and costs are higher because of additional transport. Strength and stiffness properties of aggregates are frequently less favourable when compared to gravel. Durability aspects, like freeze-thaw resistance, can also limit recycled aggregate use. Improved demolition and crushing techniques may reduce the price for aggregate in the future. If more crushing plants become operational, transport costs will also go down.

Industrial residue materials like thermally manufactured aggregates from pulverised fuel ash and phosphor slag can, of course, also replace natural aggregates.

Raw Material: Availability and Replacement

As stated before, raw material supply is coming under increasing pressure. The extraction of large quantities of marl, sand, and gravel frequently entails drastic changes in the landscape. Hence, the Dutch government pursues a policy of severe reduction of opencast mining of minerals. This in view of conserving eco-systems and scenic values and of saving the scarce space otherwise taken up by mining. In principle, authorisation for the mining of these materials is only granted if there exists a solid area redevelopment plan.

Thus economic and ecological reasons have led to the search for alternative raw materials with the aim of saving raw materials and energy.

Marl

Secondary raw materials are currently used on a large scale as a marl replacement. These materials are mainly blast furnace slag and selected pulverised fuel ashes from power stations.

Cement manufactured with portland cement clinker is referred to as portland cement. Portland cement is the most frequently used cement in the entire world. When replacing 25 % of this portland cement clinker by pulverised fuel ash, we obtain portland fly-ash cement. The pulverised fuel ash is a residue material from pulverised carbon-fired power stations. Thirty-five to seventy per cent of blast-furnace cement consists of blast-furnace slag sand, which is a residue material from pig iron manufacture in blast furnaces.

Portland fly-ash cement and blast-furnace cement are cement types containing little marl in which residue materials have been put to high-quality use. Large quantities of blast-furnace cement and portland fly-ash cement are used in the Netherlands.

The most important by-products of clinker manufacture are flint produced during marl mining and slag iron and firestone derived from blast-furnace slag. These three materials (160,000 tons in all per annum) are again used in road construction as a sub-base material.

Gravel and sand

Sand and gravel are won using the hydraulic methods of suction and dredging. The sand, gravel, and clay mixture is separated out into fractions and subsequently washed and decontaminated. Sea sand is won in the same manner.

The sand and gravel supplies are large enough to meet the needs for centuries to come. A disadvantage being that this opencast mineral mining requires a great deal of space, viz. 0.4 to 0.5 km^2 per annum. A third of this is redeveloped into land and the remainder into gravel ponds.

Gravel

Large-scale gravel mining contributes to lowering the groundwater level. This in turn leads to scorching and damage to the eco-system. The extraction of crushed rock, like granite and basalt, results in damage to nature and landscape.

Because of the impending gravel scarcity, demand for gravel-replacement materials is rising. Crushed hard rock and secondary materials are suitable gravel replacements. Natural rocks can be used in the crushed form as gravel replacements in concrete. An example of this is hard limestone used in Belgium and Germany and granite in Scotland and Norway. Yet the extraction of these rock types also disrupt the landscape. In addition, these materials are shipped over great distances. The gravel manufacturing process is confined to dredging, washing, and sieving. Natural rock still requires crushing, resulting in additional energy and noise nuisance. Natural rock also contains a higher radioactive substance content than gravel, resulting in increased radiation risk.

Sand

The sand extraction process and its impact on the environment are similar to that of gravel extraction. Sand supplies, however, by far exceed gravel supplies and the sand extraction areas are of a lesser scenic value. A major part of the sand is won from rivers. Even sea sand can be used if desalinated. Sea sand is usually a fine, graded, and somewhat granular sand that requires mixing with land sand. If long-haul transport is required for sea sand, the environmental benefit of land sand frequently exceeds that of sea sand.

Energy Consumption

The cement industry and governments of several countries have concluded agreements on reducing the use of energy.

Clinker firing (approx. 70 %) mainly determines energy consumption during cement manufacture in the Netherlands. Raw material and clinker grinding take up 10 % and 20 %, respectively, of the total energy requirement. The energy content of portland cement is 4.2GJ/ton, for portland fly-ash cement it is 3.6 GJ/ton, and for blast-furnace cement it is 2.5 GJ/ton.

Energy consumption in Dutch cement manufacture has fallen considerably owing to a number of measures. These measures are: Use of the wet process has been stopped since 1989 and the dry process, that requires 35 % less energy, has been applied exclusively; A more constant incineration process that requires 5 % less energy can be achieved by improved furnace control systems and fuel facilities; Primary fuels can be saved by using secondary fuels; The roller press can be used for cement grinding, thus reducing energy consumption. Energy consumption in industrial concrete product manufacture is low.

Fresh concrete manufacture takes place by mixing the required quantities of cement, sand, gravel, water, and possibly fillers and admixtures. Process energy is low: 0.004 GJ/ton of fresh concrete.

Energy consumption during concrete aggregate manufacture is approximately 0.26 GJ/ton of aggregate, amounting to 60 % of the energy content of primary concrete.

The energy content of cement is predominantly determined by the portland cement clinker content. Portland cement has a considerably higher energy content than blast-furnace cement, whereas the energy content of portland fly ash cement lies between the two.

There are two portland cement clinker-manufacturing processes: the wet and the dry process. Raw material grinding and mixing is easier in the wet process, yet requires 40 % more energy than the dry process. In the southern Netherlands, marl is extracted at increasingly greater depths and is thus wetter, whereby drying requires additional energy. The furnace process by far uses up most energy, notably 69 % owing to the extremely high furnace temperatures. Raw material and fuel preparation takes up less than 10 % of the total energy consumption. The grinding process requires 18 %. In all, the manufacture of 1 ton of portland cement takes up 3.6 GJ of energy. Blast furnace cement and portland fly-ash cement requires less energy. Over the past years, energy consumption has been considerably reduced owing to energy-saving measures.

Energy consumption in fresh concrete plants is low; thus drastic savings are not necessary. The reuse of waste materials like rinsing water and residue concrete can, however, result in savings.

Release of Harmful Substances During Manufacture

Cement manufacture

To maximally reduce the harmful effects of marl extraction on nature and landscape, the cement industry has taken a series of measures. The most important emissions released during marl extraction and cement manufacture are dust and NO_x. Over the past years, the Dutch cement industry has been able to reduce dust-, NO_x-, and, SO_2-emissions during clinker and cement manufacture.

Dust

Dust is released when feeding gypsum and portland cement clinker, during storage, in the dryer, in the grinder, and during cement discharge. Cement plants are well equipped for dust emission reduction. Dust is produced chiefly during fuel preparation and cement handling. Dust emission remains below the standards established by the Dutch Authorities.

Dust is produced during marl transport, storage, transfer, and crushing. Processing operations in the plant, like marl grinding, portland cement clinker making, and cement grinding, also produces dust. Over the past years, total dust emission in the Netherlands has been considerably reduced, viz. from 335 tons per annum in 1983 to 150 ton per annum in 1995.

Practically all dust is captured in electrostatic filters and filter cloths and fed back into the process. Maintenance personnel of the cement mixer and the packing plants are most exposed to cement dust. Nevertheless, concentrations at all times remain below 10 mg/m^3, the maximum permissible concentration for inert matter. Based on the dust concentrations found, it has been established that the fly-ash dust concentration amounts to maximum 2 mg/m^3 and the heavy-metal concentration is but a fraction of the maximum permissible concentration.

When making portland cement clinker, fluor compounds and a small quantity of heavy metals are released in addition to flue gases and dust.

In cement plants using blast-furnace slag as a raw material for making cement, flue gas emission remains relatively limited. This is because of the total energy principle and the use of 'clean' fuel, viz. natural gas.

Dust emission can occur during the feed of raw materials for fresh concrete making, as well as during mixing. The dust quantity emitted does, however, depend on to what extent closed-loop systems are employed. Prefab industry emissions are low.

Table 1 Emissions during clinker and cement manufacture (ENCI, 1990)

COMPONENT	EMISSIONS DURING CLINKER MANUFACTURE (kg/TON OF CLINCKER)	EMISSION DURING CEMENT MANUFACTURE (kg/TON OF CEMENT)
CO_2	408	33
NO_X	3.0	0.09
CO	0.50	0
Dust	0.21	0.12
SO_2	0.09	0
F	0.0002	0
Heavy metals	0.00005	0

SO_2 and NO_x

Cement manufacture produces harmful substances and waste. Favourable conditions prevail in the clinker furnace; thus relatively polluted fuels can still be used in a sound manner. Elevated furnace temperatures make sure the organic components are completely broken down. Heavy metals are bound in the clinker. The most important emissions are the CO_2 from flue gases and decarbonation and the NO_x from the furnace itself.

Primary fuels like lignite and secondary fuels can be employed because of the elevated flame temperature in the clinker furnace, the extended residence time, and the large surface area of the lime. The SO_2 produced during the highly sulphurous lignite incineration is mainly bound to the abundantly available lime. A disadvantage being, however, that a lot of NO_2 is realised at these elevated temperatures. Hence, the Dutch cement industry has made this an area for special attention.

The most important emissions during sand and gravel extraction are SO_2, soot, and hydrocarbons resulting from the diesel fuel. Gravel, sand, and marl are transported over relatively great distances. Although the main part is shipped, there is question of harmful environmental effects. Dust emission only occurs when transferring sand and gravel from the ship.

Table 2 CO_2 emissions of cement products

PRODUCT	QUANTITY	CO_2/g	PRODUCT	QUANTITY	CO_2/g
Portland cement 32.5R	1 kg	927.75	*Portland fly-ash cement 32.5R*	1 kg	708.93
-Portland clinker	0.937 kg	887	-Pulverised fuel ash	0.261	0
-Natural gypsum	0.0315 kg	3.75	-Portland clinker	0.701	664
-Synthetic HF anhydrite	0.0315 kg	1.26	-Natural gypsum	0.019	1.39
-Electricity	0.043 kWh	34.3	-Synthetic HF anhydrite	0019	0.757
-Barge	0.03736 tkm	1.44	-Electricity	0.0526 kWh	41.9
			-Barge	0.023 tkm	0.884
Portland cement 42.5R	1 kg	945.11	*Blast-furnace cement 42.5R*	1 kg	304.22
-Portland clinker	0.937 kg	887	-Blast-furnace slag	0.797 kg	0
-Natural gypsum	0.0315 kg	2.31	-Portland clinker	0.233 kg	221
-Synthetic HF anhydrite	0.0315 kg	1.26	-Natural gypsum	0.042 kg	3.07
-Electricity	0.0666 kWh	53.1	-Synthetic HF anhydrite	0.007 kg	0.28
-Barge	0.03736 tkm	1.44	-Electricity	0.061 kWh	48.6
			-Barge	0.067	2.57
			-Gas energy	048 MJ	28.7
Portland cement 52.5R	1 kg	979.94	*Blast-furnace cement 42.5R+*	1 kg	363.54
-Portland clinker	0.945 kg	865	-Blast-furnace slag	0.799 kg	0
-Natural gypsum	0.027 kg	1.98	-Portland clinker	0.277 kg	262
-Synthetic HF anhydrite	0.027 kg	1.08	-Natural gypsum	0.042 kg	3.07
-Electricity	0.0999 kWh	79.6	-Synthetic HF anhydrite	0.007 kg	0.28
-Barge	0.03336 tkm	1.28	-Electricity	0.081 kWh	64.6
			-Barge	0.57 tkm	2.19
			-Gas energy	0.525 MJ	31.4
Masonry cement	1 kg	513.64			
-Portland clinker	0.489 kg	463			
-Limestone/marl	0.471 kg	1.3			
-Natural gypsum	0.02 kg	1.46			
-Synthetic HF anhydrite	0.02 kg	0.797			
-Electricity	0.056 kWh	46.2			
-Barge	0.023 tkm	0.884			

When demolishing concrete, emissions are mainly dust, SO_2, soot, and hydrocarbons due to the diesel fuel used in the various machines. A large quantity of dust is particularly emitted when upgrading concrete into aggregate.

The most important waste streams are firestone and slag iron that can be reused. This leaves 'ordinary' corporate waste and household waste water.

Cooling-water and household wastewater are discharged on the surface water. Waste is also produced in the form of 'ordinary' corporate waste, slag iron, slag lumps, and manufacturing residue. The largest part hereof is used as a residue material.

Concrete mortar manufacture

Concrete mortar plants are currently taking measures to obtain a better environment. These measures are mandatory and involve permits. The companies themselves also practice internal environmental protection. Employees are barely exposed to cement dust since concrete mortar manufacture has been mechanised and automated.

In addition to gravel materials and water, a small quantity of thermal energy is used when producing concrete mortar. Thus hardly any flue gases are produced in concrete mortar plants. Material residue returned from building sites and resulting from rinsing the equipment, can for the most part be recycled. The dust, the rinsing water, as well as the solid residue are harmless, although the dust can cause visual nuisance. Fresh concrete is indeed shipped at low costs, but SO_2 -emission due to the use of the highly sulphurous diesel is considerable.

Concrete mortar requires mixing water: 150 to 170 litres of water for each m^3 of concrete. The mixing water must meet the requirements with regard to chloride content, sulphate content, and acidity. Rinsing water from concrete mortar plants can be reused as mixing water. The rinsing water contains concrete mortar residue after having been used to clean the truck mixers and is subsequently poured in a settling pit. The heavier components like gravel, sand, and cement sludge settle in this pit. The rinsing residue is collected in a settling tray. Water, gravel, and cement sludge can then again be fed into the manufacturing process.

Admixtures

Admixtures are added to the concrete mortar in very small quantities so as obtain special properties during processing. Since the production lines have been automated, workers no longer come into direct contact with these substances. Admixtures, like plasticizers or air-entraining agents, are generally fairly harmless polymers that can be broken down easily. They are used to influence the properties of non-hardened concrete mortar. Admixtures are used in small quantities: usually less than 1 % of the cement weight.

The environmental benefit of plasticizers is that it makes the concrete mortar more plastic or lowers the water/cement factor whilst maintaining workability, thus giving the mortar a lower energy content.

Air-entraining agents improve workability, raise freeze resistance, and increase the watertightness of concrete mortar.

Cast in situ concrete

It is less easy to save materials in cast in situ concrete than in prefabrication. Also, there is a high probability of formwork removal oil ending up in the environment. Flue gases like soot, organic hydrocarbons, and sulphur dioxide are produced when using energy for transport, casting, and compaction. The most important waste streams are concrete mortar residue, formwork materials, and steel waste.

Cast in situ concrete means monolithic structures, i.e. without the joints and connections characteristic of prefab construction. Heavy, solid structures lend themselves more to cast in situ concrete than to prefabrication. Nevertheless, material-saving techniques, like prestressing and open ducts, can also be employed in these monolithic structures. Cast in situ concrete, however, requires more formwork materials.

Concrete products

Concrete product manufacture requires little energy and produces little harmful substances. Concrete residue does remain, but is largely recycled. Casting, vibration, and formwork removal is done electrically.

Waste consists of concrete residue, rinsing waters, rinsing residue, formwork materials, packing waste, and sweepings. The remaining residue concrete is largely recycled.

There are tools that place a load on the environment, like formwork moulds, formwork oil, insulation, and packings. The moulds are reused many times. The packings can also be used again and can moreover be made with recycled materials.

Polystyrene foam is occasionally used as formwork or as insulation. Polystyrene can also be part of the concrete product.

Formwork removal agents

There are various kinds of formwork removal agents that all contain additives and are harmful to the environment. Formwork removal agents are responsible for 15 % of the hydrocarbon emissions. Formwork oil can contaminate the soil. These days, more and more formwork removal agents are manufactured that are less harmful to the environment because they can be broken down and contain less harmful solvents and additives.

Six million litres of formwork removal agent is used in the Netherlands per annum. This means that approximately 5,000 tons of hydrocarbons end up in the environment every year and thus contribute to the greenhouse effect.

The environmental load of formwork removal agents has been much reduced over the past few years as a result of the removal of PACs (polycyclic aromatic hydrocarbons) from mineral oils, the development of decomposable synthetic products, and the replacement of environmentally damaging solvents by oil-in-water emulsions. The agents now are usually free from aromatics and no longer contain sodium nitrite used as a corrosion inhibitor. Vegetable oil-based formwork removal agents are also commercially available.

Prefabrication

Prefab element transport and assembly require energy. Harmful substances are produced because of the diesel used for transport, pile drivers, and travelling cranes and because of finishes like cement and foam. Moreover, concrete residue, chemical waste, and packing waste remain.

So as to allow for differences in material expansion, the joints are cemented together. The seams between the prefab elements are also filled with cement. This cement places a load on the environment.

Prefabrication is an industrial process governed by quality systems. Rejected products and residue from cut-aways are recycled into aggregates. Residue from cut, drilled, and sawn cut-aways on the building site either ends up in the rubble container or is returned to the mill to be crushed into aggregates.

Ionic radiation in the residential environment

Radon in the indoor air is one of the most important sources of a potent radiation dose. Radon is a radioactive inert gas that is produced by or emanated from construction materials and from the soil. This gas can be found in the outdoor air (originating from the soil), but in higher concentrations in the indoor air, notably in the crawl space. Thus, the radon concentrations in the home are determined by radon emanated from the soil and from construction materials, on the one hand, and by the ventilation airs in and moving through the house, on the other.

Since the seventies, a rising trend in radon concentrations in the home has been recorded. The Dutch Government has in the meantime developed a policy for the introduction of a Radiation Performance Standard with associated limiting values. This standard wil serve to compute the total dose of radioactivity sustained annually by staying in houses or residential buildings.

Residue Materials in Concrete

Processing residue materials in concrete already is a commonly used form of reuse. Examples are: partial replacement of cement by fly ash, of gravel by various types of slag, and of fillers by fly ash or furnace bottom-ash. For sound use of residue materials in concrete construction materials, the focus should be on the environmental and labour protection aspects of manufacture, use, and the waste stage. It is important to consider whether situations can occur in the different stages that pose a threat to health due to the release of toxic and carcinogenic substances.

The leaching behaviour of concrete in which residue materials have been processed, has been investigated for many types of residue materials. Mostly, the cement or the aggregate in the concrete is fully or partly replaced by the residue material. Examples of residue materials used are: fly ash, furnace bottom ash, Lytag, aggregates, slag from waste incinerators, phosphor slag, jarosite end slag, ground blast-furnace slag, and LD steel slag.

Concrete used as a construction material is a suitable means for re-using residue materials because of low permeability and high alkalinity which allows for keeping the environmental attack of most components within acceptable limits. Yet, the leaching behaviour of concrete can be affected by factors like permeability, cement type and cement content, age, and climatological conditions.

Leaching behaviour is not the same for each component for that matter. Diffusion-controlled behaviour occurs in a large number of components.

Permeability

Concrete permeability, which is determined by the raw materials and the manufacturing method used, of course directly affects leaching behaviour. In the worst case scenario, convective transport might even occur, for instance, in concrete (even of low permeability) that is subject to pressure loss or to capillary action. The cement type (blast furnace versus portland cement) can affect permeability and consequently leaching behaviour.

Ageing

Concrete age can also affect leaching behaviour. Chemical changes in the concrete will affect leaching behaviour from the time of manufacturing. The high alkalinity in concrete will cause the various residue material components to immobilise as a result of precipitation (hydroxides and carbonates). This will also lead to the creation of areas of poor permeability in the concrete. Because of hardened cement paste formation with time, an increasingly dense concrete structure will come about, possibly reinforced by the self-bonds of the fly-ash residue material. This will additionally lower permeability. Finally, concrete weathering can occur in practice and the residue material can be subjected to changes with time, such as hardening (recrystallization).

The formation of other salts (crystalline structures) can improve or deteriorate solubility, resulting in a similar change in leaching behaviour. Concrete carbonation due to CO_2 from the environment severely lowers concrete alkalinity. This will cause various precipitates to again dissolve, whereby the contaminations become more available for leaching.

Particularly contact with the somewhat acid rainwater is fatal for carbonated concrete. The carbonate is dissolved and the immobilised components change to the water phase. There is then no longer question of diffusion-controlled behaviour, but rather of solubility and washing away. Nor can the material then still be referred to as 'durable'. Temperature not only affects concrete durability, but also directly affects leaching behaviour. The mobility of the various components is lower at an average temperature of 10° C than at 40° C, thus creating a difference in the degree of leaching. Hence, diffusion-controlled leaching behaviour depends on the temperature. In addition, there may be changes in the various components' solubility due to temperature variations. Leaching behaviour can vary for different types of components.

Leaching behaviour of concrete in which residue materials have been processed is to a high extent determined by the chemical and mineralogical composition that can change with time, even when dealing with a durable product.

In most cases, concrete leaching behaviour develops according to a diffusion-controlled process.

Soil Protection Measures

Many industries deploy activities that hold a potential risk of soil pollution. The Government imposes severe requirements on the facilities that serve to prevent soil pollution. Liquid-proof soil protection facilities are mandatory for risky industrial activity. The Environmental Management Act and the Soil Protection Act constitute the legal framework.

Concrete can be used in environmental protection structures, like liquid-proof pavings in, for instance, petrol stations and in the agricultural and chemical industry. It can also be used in nutrification-stores and -silos and for waste material storage (CO_2-deposits). Hence concrete is frequently used in soil protection facilities, both as prefab elements and as cast in situ.

Recent research has uncovered the most important mechanism of liquids penetrating into concrete as well as the underlying theory. This penetration is almost completely determined by capillary absorption. Insight into the relation between the liquid and the most important aspects of concrete technology is good. The denser (finer) the pore structure of concrete, the slower or lesser the liquid penetration. Viscosity and surface pressure of the liquid, and notably the relation between these two properties, are normative for the rate of penetration.

The capillary absorption test is used to determine liquid penetration in concrete. This method of testing can be used to compute the design thickness of a liquid-proof soil protection concrete structure in a given situation. The method can also be employed to determine an existing facility's remaining lifetime during which the requirements of liquid-proofness are still met.

Life Cycle Analysis (LCA)

An environment-oriented life cycle analysis portrays environmental loading during a material's entire life cycle, i.e. from raw material extraction up to and including demolition and reuse. This will result in an environmental profile, which can be used to assess the environmental impact of products or activities. LCA can also be conducted to improve a product or a manufacturing process. LCA is then employed to track down processes that take a relevant part in the overall environmental load and is referred to as improvement analysis.

The environmental effects during the entire life are inventoried so as to obtain a good idea of a concrete structure's environmental load. If products are again used as a secondary raw material after the waste stage, we speak of a life cycle. An environment-oriented life cycle analysis can be conducted in a qualitative and a quantitative manner. International agreements have been concluded on how to conduct a quantitative life cycle analysis, referred to as the LCA-study.

LCA involves a number of stages. In the targeting stage, the target of the study and the functional unit used are established, for example: one square metre of a storey-floor spanning a certain distance.

In the inventory stage, the processes occurring during the life cycle and the ensuing environmental interferences are examined. Environmental interferences can be, for instance, raw material and fuel extraction and emission and waste production. Once all processes and environmental interferences have been inventoried, the contribution of the various environmental interferences to the environmental aspects is examined in the classification stage.

(Raw) materials are examined for each product. The contributions of all environmental interferences are listed for each environmental aspect. These results in a list of materials extracted from the environment and of emissions into the environment for all manufacture-related processes necessary for the performance of a product, a service or an activity during the entire life cycle. This is referred to as the final environmental profile.

The environmental profile is interpreted in the assessment stage. It is described how to wind up a life cycle analysis. Suggestions are put forward on how to present the results and directions are given for the improvement analysis.

Simapro

LCA-studies can, in principle, be conducted for all material and functional unit combinations. A specially adapted programme on the use of concrete in construction is being developed under the name SIMAPRO. SIMAPRO is a computer programme that can be used to process and interpret the requisite data.

SIMAPRO's most recent version contains a comprehensive series of base data relevant to manufacture, design, and use of concrete products and structures. The environmental data have been updated.

REFERENCES

1. GLAVIND, M et al Green concrete for the future, Cammet, 1998

2. HAUGAARD, M et al Industry analysis, concrete - Cleaner technology in concrete Production, Danish Environmental Production Agency, 1997

3. MOELLER, J S Denmark's Great Belt Link, ASCE, Annual convention 1989, New Orleans

4. POULSEN, K M Fehmern Baelt forbindelsen, Indragelse af globale miløaspekter, Thesis (in Danish), Technical University of Denmark, 1999

5. SAID, R The Aswan high Dam revisited. IABSE, 15th Congress, 1996, Copenhagen

6. HARTOEFT, L Q AND MEYER, F. Bridge management today and for the future

7. Stream analysis of construction materials. University of Technology Delft, 1999

8. HENDRIKS, Ch F, Durable and Sustainable Constructions Materials, The Building Cycle, Aeneas, Best, The Netherlands, in press (1999)

INNOVATION IN CONCRETE CONSTRUCTION PROCESS

O Kornadt

Philipp Holzmann Group

Germany

ABSTRACT. The innovation in the concrete construction process is driven by developments in material, information and automatisation technology. The path to advanced concrete construction process is shown in the paper by describing both realised innovative projects and research results on the threshold to realisation.

Concrete will remain in the future the principal material used in the construction industry, and the areas of application will continue to grow and adapt to the needs of an ever changing society.

Keywords: High performance concrete, Self Compacting concrete, Recycling concrete, SIFCON, Hybrid construction, Bubble-deck, CAM in concrete, Pre-casting

Dr Oliver Kornadt studied physics, mathematics and economics in Saarbrucken and Aachen. After some years working at the nuclear research centre CERN in Geneva he obtained his PhD. In the following years, he worked as research engineer in the energy industry. Today he is head of the department for co-ordination of research and development and responsible for the area of building physics within the Philipp Holzmann Group. In 1998, Oliver was elected President of ENCORD (European Network of Construction Companies for Research and Development), a permanent grouping of several leading European construction companies.

INTRODUCTION

The drive that engineers have in forever looking for new and better solutions as well as the most suitable technical advancements influences the development and innovations made in the concrete construction process.

Over time concrete has drastically developed as a construction material. The advancements in methods of construction of concrete structures lie in the further developments of concrete as a material, the environmental demands, the development of processes in concrete and composite construction, innovation in the production if concrete elements and structures and the futuristic concepts of design teams.

FURTHER DEVELOPMENTS IN CONCRETE AS A MATERIAL

High-Performance Concrete

The well tried and tested normal strength concrete has played, and will continue playing, its part in considerable increasing the functional range and mechanical characteristics of the structures we produce - but normal strength concrete is only the starting point of that which can be achieved.

One further step from this is high-strength concrete, which opens up many more possibilities for construction methods. High-performance concrete is an even further step up the ladder in helping us to achieve the forms and heights of the structures required to meet the needs of modern day technology and society.

The developments in high-strength concrete result in a constructive and economical advantage in the construction of concrete high-rise buildings in comparison to those with a steel skeleton [1,2]. In 1990 J.A. Jones the U.S. subsidiary of Philipp Holzmann completed work on the 65 floors office building "311 South Wacker Drive" which, with a height of 292 m, remains to-day the highest concrete building in the world, see Figure 1.

The experience won in the USA is today used in European projects where slender columns with high compressive strength are needed in order to reduce their used area. For example, nominal strength concrete of 85 N/mm^2 was used in the 186 m high bank building, the "Trianon", in Frankfurt am Main, see Figure 2.

The development of the strength class continued from 85 N/mm^2 up to strengths of 115 N/mm^2.

At first an increase in the strength of concrete was only possible by lowering the water / cement ratio (w/c ratio). This was possible through improvements in the concrete mix, especially in the areas of admixtures and plasticisers, the limit was however reached at a w/c ratio of 0.40.

Figure 1 "South Wacker Drive", Chicago, USA. The highest concrete sky scraper in the world

Figure 2 The „Trianon", Frankfurt Germany, with a height of 186 m

It is not possible to achieve a further increase in the nominal strength class of over 80 N/mm^2 when using the classical raw materials. or even when using w/c ratios under the 0.30 or 0.35 mark.

In order to leap forward into the realm of even higher strength concrete the addition of silicate dust is necessary, a by-product of the production of ferrochromium silicate. Using w/c values between 0.20 and 0.30 and through the addition of silicate dust the compressive strength of the concrete can be increased to values of more than 100 N/mm^2

Disorders in the bond zone between the aggregate and the cement grains occur in normal concrete even when low w/c ratios are used. These disorders are caused by calcium hydroxide, ettringit and water, and they affect the concrete strength.

This problem is drastically reduced by the addition of silicate, which is 100-times finer then cement, and reacts pozzolanically to form, together with the calcium hydroxide, calcium silica hydrate. This improves the bond between the aggregate and the cement grains. Disorders in the bond occur only in a very minimal extent. Moreover, due to the presence of the calcium silica hydrate the mortar matrix becomes denser and firmer. The concrete structure is overall more homogeneous.

The scanning electron microscope makes it possible to have an insight into the micro structure of the concrete and to produce image representations of the process. The difference between the texture structure of the mortar matrices can be seen in Figure 3 and 4, and this results in the differing density and firmness of the concrete types.

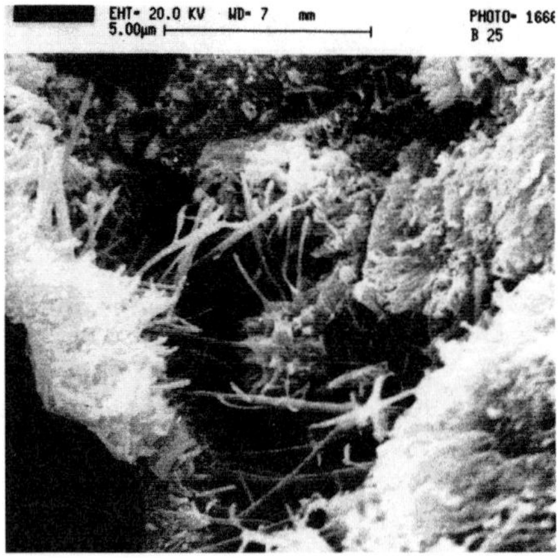

Figure 3 Micro structure of normal strength

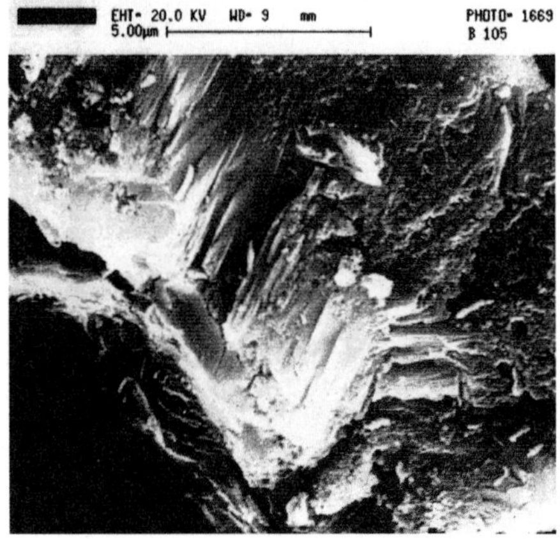

Figure 4 Micro structure of high strength concrete, 105 N/mm^2

In normal concrete with w/c ratios of 0.60 a 25 volume percent of capillaries exists in the mortar matrix, into which the individual hydration products can grow. In the case of high strength concrete, with w/c ratios of less than 0.30, there is a much lower volume of capillaries pores. therefore the cement grain structure is denser.

Tests have shown that, whereas in normal concrete the weak spot, or cracking, occurs in the mortar matrix, in high strength concrete failure lines pass directly through the coarse aggregate, see Figure 5 and 6 [3]

Based on these findings the Congress Centre, in the Frankfurt trade fair grounds, was constructed in l 99S using concrete with a nominal strength of l OS N/mm^2. For the use of the high-strength concrete it was necessary here, as was with the "Trianon" building, to have a complete and practice suited quality assurance system.

For all such projects carried out to date the increase of the concrete strength was the highlight objective.

In the construction of the Japan Centre as shown in Figure 7, in the centre of Frankfurt, high strength concrete was, for the first time in Germany, used not only for the columns subjected to high compressive forces but also for the structural elements subjected mainly to large bending moments.

The Japan Centre, with a height of 115 m above ground level, is no sky scraper, but due to the open face facade high strength requirements are imposed on the concrete elements, see Figure 8. The slender frame connection points, due to the open facade design, have to withstand a load combination of normal, shear and bending loading.

Figure 5 Failure face of normal concrete test sample, failure occurs in the mortar matrix

Figure 6 Failure face of high strength concrete test sample, failure occurs in aggregate

Only through the use of high strength concrete it was possible to achieve the slender cross-section dimensions for the frames vertical and horizontal members necessary for the large window fronts, see Figure 9. For the use of the 105 N/mm^2 concrete, subjected to bending stresses, it was necessary to have a individual-case technical approval from the standards authorities.

The development in enhancing the compressive strength is going on. It is possible to obtain strengths up to 200 N/mm2 and this of course will change the concrete construction process even further. Figure 10 shows four cross sections, the two cross sections on the right are constructed from normal strength concrete, the second from the left from structural steel and the left cross section is constructed from super high strength concrete. All the sections have the same load carrying capacities. It can be recognised immediately that both the envelope and the self weight of a structure can be reduced through the use of super high strength concrete.

Self Compacting Concrete

The use of self compacting concrete will lead to a more industrialised production, significantly reduce the technical cost of in-situ cast concrete construction, improve the quality, durability and reliability of concrete structures and improve health and safety on and around the construction site.

Through research and development projects, methods have been developed which solve the problems of aggregate segregation, see Figure 11.

Figure 7 The Japan Centre, Frankfurt

Figure 8 The open-face facade of the Japan Centre, Frankfurt

Figure 9 High-strength slender columns, Japan Centre, Frankfurt

Figure 10 All cross sections have the same load carrying capacity.
Left: super high strength concrete, Right: Convention concrete

Results from tests show that it is possible to produce a self compacting concrete, both with and without steel fibres. When the complete technical and economical specifications required for the use of self compacting concrete are attained, and the cost of plasticisers is reduced, it can be expected that the use of self compacting concrete will increase.

Initial the aim, in full-scale housing tests which was carried out, was only to cast some of the walls and slabs with self compacting concrete, but when the site personnel had experienced the advantages of the new technology they did not want to return to the use of traditional concrete. Through the elimination of the use of vibrators the working environment on-site was considerable improved, also a lot of the heavier work and the noise from the vibrators was eliminated. Due to the positive reaction and results it was decided to cast the whole office complex with self compacting concrete, which meant a volume of about 3000 m^3.

Around 433,000 m^3 of self-compacting concrete have been placed in the anchorage's of Akashi-Kaikyo Bridge, which is the world's longest suspension bridge. This is one of the first examples of self-compacting concrete being used in such a large structure. One reason for its use was to shorten construction time. Up to 1900 m^3 of concrete was placed per day. To improve placeability, self-compacting concrete with maximum coarse aggregate size of 40 mm was developed and used, see Figure 12.

ECOLOGICAL POINT OF VIEW

Recycling Concrete

The estimated amount of demolition waste is expected to rise dramatically in the future, see Figure 13. The recycling of this waste is of up-most necessity.

Figure 11 Slump flow of self compacting concrete

Figure 12 The Akashi-Kaikup Bridge, the world's longest suspension bridge

This will result in the development of an intelligent and efficient material economical plan in the construction industry [4].

In the standard method of demolition, as shown in Figure 14, the resulting waste is a mixture of the various materials from all the building components and therefore, not only do difficulties arise in recycling the waste, but the costs of such works are not justifiable [5]. This resulting waste mixture is not only due to the method of demolition, but also the method of construction, as the need for recycling of construction materials was not recognised at the time of construction of the now old buildings.

The construction industry is presently working on developing environmentally friendlier structures which are also calculated to be recyclable on the grounds of their construction method. Questions been addressed include the required alterations in the planning of new structures, in both design and construction phase, that at the end of its life-time the structure not only acts as a source of reusable raw materials for further projects, but also has a minimum of non-reusable materials. In such matters the areas of energy and grey-energy must also be addressed. Also, a reduction in the amount of different building materials results in an easier recycling process.

The increase in the life-time of a building results in environmentally friendliness. Therefore the structural and finishing works should be conFigured in such a way that a change of use of a structure can be carried out with minimum repercussions and, if possible, without changing the load-carrying structure [6].

The manner and extent to which materials can be recycled depends on the manner in which the structure is demolished. This means that in the future demolition processes will require more input from the engineering side and also a better planning process, see Figure 15. A point in time will come where a quick demolition with no resulting materials for reuse will not be permitted. Certain regional governments only grant demolition permission to projects showing a reduction in waste.

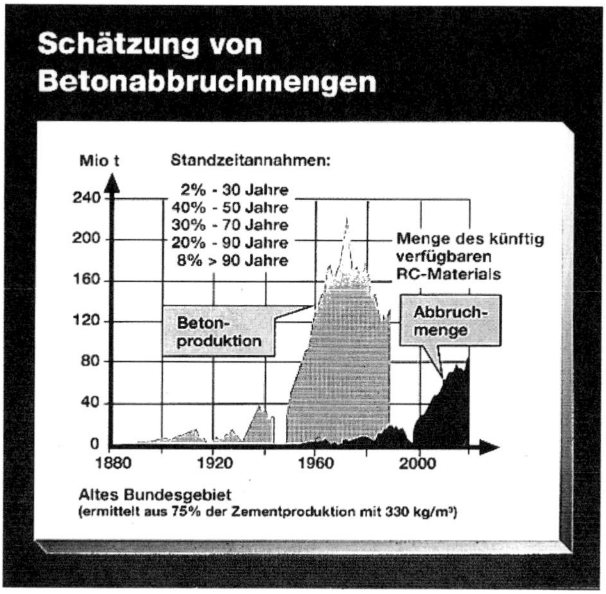

Figure 13 Recycling quantities will increase dramatically
Betonproduktion = Concrete production
Abbruchmenge = demolition waste

Figure 14 Standard method of demolition, resulting waste is a mixture of all building components

Figure 15 A well planned demolition process results in a higher recycling quantities

Figure 16 Crushed stone produced from recycled concrete waste

When examining the typical high-rise construction most, if not all, the waste materials can be recycled or downcycled (reused in a lower value class). Figure 16 shows crushed stone produced from recycled concrete waste.

Figure 17 shows two core cross-sections from a test series, on the right is a core with natural aggregate and on the left a core with recycled concrete as aggregate. The recycled concrete originates from a demolished building which had being constructed in 1962 with concrete with a nominal strength of 30 N/mm^2. The structural elements were separated from the finishing works, delivered to a recycling plant, and using a jaw-breaker, sieving machine and a magnate separator, were broken and sieved. The remaining foreign substance were sorted out per hand. The results showed that for core samples with constant consistency, controlled by addition of water, the values of the compression strength, the tensile strength in bending and the elastic module dropped, as had been expected, with increasing amounts of concrete chippings. Through the use of plasticisers for controlling the consistency of the mix no effects to the concrete characteristics were shown due to the use of recycled materials. Only the elastic module sank in value for test cores using 100 % recycled aggregate to approx. 77 % of that of concrete with 100 % natural aggregate. The tests confirmed that, assuming sufficient quality control, and using a coarse aggregate through to 100% concrete crushed stone a workable concrete with a nominal strength of 3 5 N/mm^2 can be produced.

One critical issue in the reuse of concrete waste are the government regulation, which vary thoughout Europe [7]. For example, in the Netherlands the reuse of concrete waste is permitted but in Germany this process is not yet standardised in the building regulations. Only through the use of a special technical approval from the standards authorities the headquarters for the German Federal Environmental Foundation, shown in Figure 18, in Osnabruck could be constructed using recycled concrete. A requirement of the Environmental Foundation was that the building structure and finishing works were to be constructed from materials with environmentally friendly characteristics. Therefore many recycled materials were use, such as cellular glass, Terrazzo from recycled glass, cellulose paper materials and of course recycled concrete.

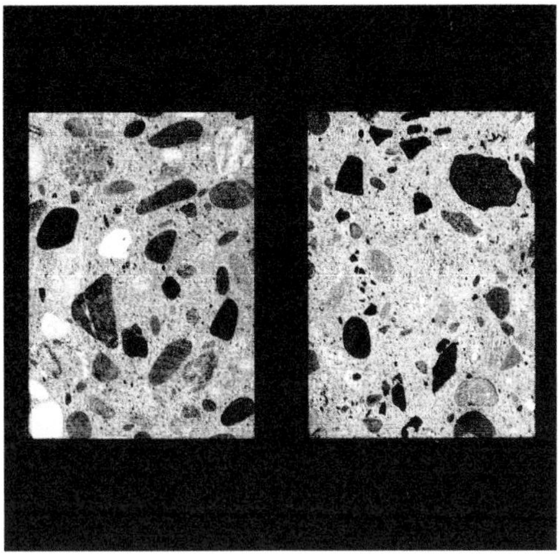

Figure 17 Sections through test samples left: 100 % natural aggregate, right: 100 % recycled concrete aggregate

Figure 18 The headquaters of the German Federal Environmental Foundation, Osnabrück, was constructed with recycled concrete aggregate

DEVELOPMENT IN THE PROCESS OF CONCRETE AND CONCRETE COMPOSITE SYSTEMS

SIFCON

The impermeability of concrete is, especially in view of environmental issues, of great importance. For example, the contamination of the ground and / or groundwater from petrol station pollution must be prevented, see Figure 19. This was, unfortunately, not taken into consideration in the past when constructing such facilities. New concrete surfaces must be able to withstand leakage and be able to store water endangering substances, thus preventing any possible contamination of the ground or groundwater [8].

For such cases high performance concrete can be used, as the impermeability of the concrete is achieved through the density of its micro structure. This is valid for both the renovation of existing facilities and the construction of new facilities in which environmentally un-friendly substances are either stored, filled or can be knocked over.

For the renovation of existing facilities, Philipp Holzmann together with the Technical University in Darmstadt, Germany, have developed the SIFCON-method. SIFCON stands for Slurry Infiltrated Fibre Concrete. In this method the reinforcement is completely replaced by a very high percentage of steel fibres. The steel fibres make up more then 10 % of the volume and therefore more then 10-times that used in typical steel fibre reinforced concrete. It is therefore possible to achieve dense layers with a much lower slab thickness, which are practically crack-free and impermeable, and also due to the high ductility the layers have a deformation behaviour similar to that of a steel plate.

With present day methods it is not possible to mix high amounts of the required steel fibres into the fresh mortar. Therefore, with the SIFCON-method, the steel fibres are distributed directly onto the cleaned underground, then covered with a thin high performance mortar and finally condensed using a vibrating screed, see Figure 20.

Tests were carried out with the dangerous substances, and it was found that none of the substances penetrated the slab during the required and specified minimum life-time of the slab.

Compression strengths between 90 and 105 N/mm^2 and tensile strength in bending between 35 and 45 N/mm^2 were achieved with SIFCON. Thus implying a high ductility. A complete filling of all the void areas in the steel fibre aggregate is guaranteed due to the composition and low viscosity of the mortar, see Figure 21.

The advantages in this method lie not only in the high load carrying abilities and the high impermeability, but also in the fact that the substructure does not need to be removed and the load carrying layer does not need to be renewed, see Figure 20.

SIFCON is not only suitable for the sealing of small areas in petrol stations, but also for new covering on runways. Many of the worlds runways are severely damaged due to the constant load cycles from landing and starts and this damage also affects the well being of the passengers.

Figure 19 Contamination of ground water under petrol station is possible as surface is not impermeable

Figure 20 SIFCON can be applied directly to existing substructure

The high steel fibre content also increases the abrasive resistance and the impact resistance in comparison to conventional concrete coverings. The maintenance requirements are also reduced. The new SIFCON layer can be applied jointlessly without upgrading the existing concrete covering. This enables restoration of runways in a very short time period. In order to cover large areas in an economical manner with the steel fibres a dispersion machine is required, and this has been developed by the German Asphalt Company, see Figure 22.

Figure 21 Failure and cross section of SIFCON test samples

Figure 22 Steel fibre dispersion machine, specially designed for SIFCON

Hybrid Construction

Many interesting developments have occurred in the area of composition construction methods. Hybrid construction describes any construction in which at least two materials are used. An example of this the use of hybrid construction for bridges, see Figure 23 and 24, showing the "Drei-Rosen" bridge in Basel. The optimal properties of each individual building material are utilised. Compressive forces are transferred through the medium of concrete, tension through tendons and shear forces in structural steel. Weds constructed from folded steel have the advantage that they attract next to no longitudinal forces. The web therefore takes-up, in comparison to conventional solid concrete cross-sectional bridge decks, a large proportion of the prestressing.

Figure 23 and 24 The "Drei-Rosen" bridge, Basel, is a hybrid construction

The use of hybrid construction for bridges has many advantages over conventional construction: the self weight of the bridge is reduced, due to the optimised cross section and much slender bridges are possible, which are, for example in the case of railway bridges, much more desirable.

A further examples of composite construction is a wood-concrete composition, see Figure 25. The wood transfers the tensile forces, equivalent to the function of reinforcement bars in conventional reinforced concrete construction, and the concrete transfers the compressive forces [9].

A series of houses have been constructed in Switzerland using wood-concrete composite roofing. Through the use of this composite construction the high noise reduction and fire rating requirement imposed on conventional wood roofing are easily fulfilled, this factor also makes the construction method very interesting in the areas of renovation of existing buildings and change of use.

BubbleDeck

The BubbleDeck is a further example of concrete innovation. This is a two-way spanning hollow slab in which plastic spheres keep the concrete from entering areas where it performs no function, see Figure 26 and 27.

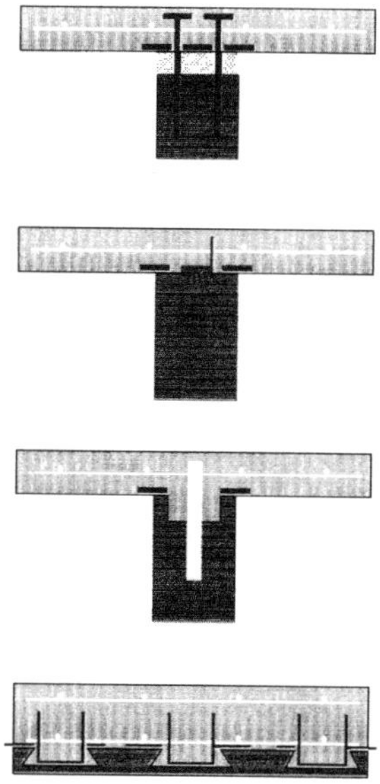

Figure 25 Wood-concrete composition constructions

By fitting the steel mats and the spheres together into the required raster, the optimal cross-section can be achieved and simultaneous there is no loss in the utilisation of the moment and shear capacity of the slab. Due to the reduction in the self weight of the slab a BubbleDeck has the same load carrying capacity with a thickness half that of a conventional slab.

Figure 26 BubbleDeck: the two-way spanning hollow slab

New aspects are possible due to this development: larger areas, without intermediate supports, lower construction heights for the same number of floors and also there is a higher quality control as the members are constructed under factory conditions.

Figure 27 Filigree slabs for BubbleDeck been delivered to site

Further advantages is the environmental friendly aspect - 100 kg of concrete is replaced by 1 kg of recycled plastic, there are less transportation requirements and the elements can be easily separated and recycled at a later demolition time.

The first larger project using the BubbleDeck technique is the Millennium Tower in Rotterdam, see Figure 28.

Figure 28 The millennium Tower, Rotterdam, the first larger project using BubbleDeck

INNOVATION IN THE MANUFACTURING OF PREFABRICATED CONCRETE UNITS AND CONCRETE CONSTRUCTION

Computer Aided Manufacturing

The concrete construction process is becoming more and more influenced by information and automatisation technology. A computer aided manufacturing (CAM) is possible from a technical print of view.

Philipp Holzmann constructed the Dusseldorfer Zollhof, an ensemble of three differently fashioned buildings, planned by the Californian architect Frank O. Gehry, in the centre of Dusseldorf. The buildings with wavy formed structures, are very curved and lie up to 1.70 m out of the perpendicular, see Figure 29.

For the construction of this buildings a new innovative construction method based on CAM was applied.

Figure 29 Düsseldorfer Zollhof, the art work from architect Frank O. Gehry

The outer skin of the structures were digitised by using a high performance software, CATIA. CATIA is a construction software which is used in the design offices in the automobile and aeroplane industries. The form of the outer skin can not be described by any mathematical geometry, it is a completely free-formed area. The design was purchased as a piece of art work, and the condition had to be fulfilled that the final structure matches exactly the digitised form.

The model could not be run on any existing DOS-based computers, and the data files were too large for the standard CAD-workstations. Due to the high costs of purchasing a CATIA workstation, another solution needed to be found.

The facade was designed at the outset of the detail design phase with the conventional architectural methods, and no changes were permitted thereafter. For the shop drawing phase the formwork contours and window openings were generated using the CATIA system. This was then transferred floor by floor to a AutoCAD system, with the aid of a high powered computer system, so as smaller workable data files were produced, see Figure 30. Structural formwork plans, with areas of approximately 3.0 x 3.5 m were then produced. The data from the formwork plans were then transferred to the automated control program of a milling machine, and the contours of the precast concrete forms were cut into styrofoam blocks; see Figure 31. Information for the triangulation points and the sorting order for site were included into the forms. The formwork moulds were then delivered to the precast yard, where they were reinforced and concreted, see Figure 32. The delivery of the precast sections to site followed a "just in time" schedule, see Figure 33 [10].

The development of this technique for precast section grants a higher degree of freedom to the architect in the designs of the future, and the direct transfer of formwork plans to an automated cutting program is a technique which will be much used in the future.

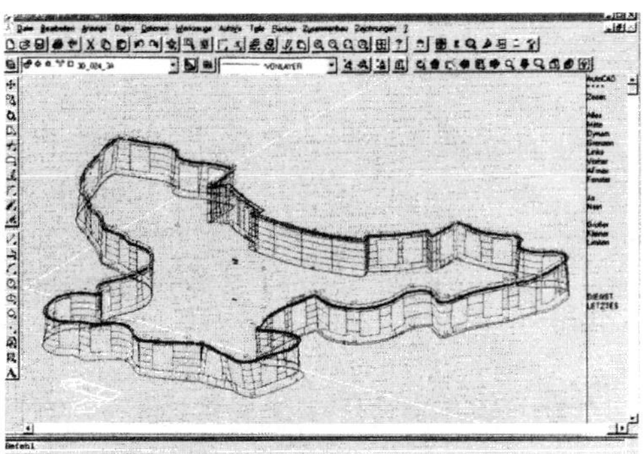

Figure 30 Floor segment in AutoCAD

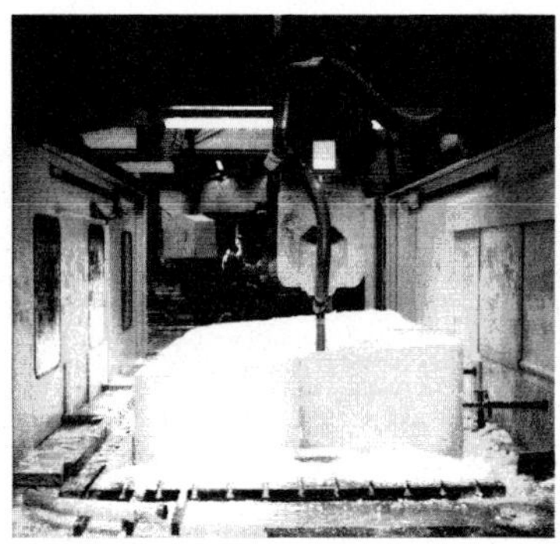

Figure 31 Automated control program sent cutting information to milling machine

Further Developments In Prefabrication

The extensive mechanisation of prefabricated concrete factories ensures that the units produced are cleaner and of a higher quality, the work process also becomes more efficient. The products emerging from the factories, such as shown in Figure 34 are both technically and aesthetically of a very high standard.

The generalisation of component building means that construction work is increasingly being transferred from the building site to a factory environment.

As far as possible, all the building's technical systems are manufactured under controlled factory conditions.

Figure 32 Work at the precast yard

Figures 33 Fixing the precast segments together on-site

The aim in component production is to develop the capacity to deliver highly individualistic project-specific orders as quickly and efficiently as possible. The attainment of the goal has been made possible by the building system compatibility principles and the automated technology.

The solutions for the load bearing components of the structural system are still tied to a basic material. Concrete, timber and steel are all used as frame materials, and each one has its own component solution. The use of different combined and composite structures is becoming more widespread. The number of alternatives is also increasing in the field of supplementary building component systems.

Figure 35 shows the installation of such a component, the technique is ideal for projects where the same room solution is repeated a number of times, e.g. in hotels. The rooms, which are completely ready when they leave the factory - including surface finishes and fixtures, can be designed and produced in the factory at the same time as the building frame is being constructed on site. Due to the considerable reductions in the construction schedule, considerable savings can be made in financing expenses.

Figure 34 Prefabricate concrete factories result in higher quality and efficiency

Figure 35 Installation of a prefabricated building component

It is, however, necessary in this form of construction that the tolerances of the building components lie in the mm zone and not in cm's, as in conventional building techniques. Furthermore, in order to relish all the advantages of this construction method the use of computer assisted planning is a must.

CONCLUSIONS

Finally I would like to show you the visions of the future - a 750 m high concrete skyscraper. Figures 36 show the sky-scraper imposed, using photomontage, on the sky-lines of Chicago and Atlanta. According to our calculations, concrete with a nominal strength of 200 N/mm^2 is needed to build such a structure.

This, however, does not have to be the limit of the visions and innovations of concrete construction process - there are many more challenges which can be tackled in the concrete construction process, as shown in Figure 37.

Figures 36 Photomontage showing a 750 m sky scraper in Chicago (left) and Atlanta (right)

Figure 37 The future ?

REFERENCES

1. KÖNIG, G AND FISCHER, J. Model Uncertainties Concerning Design Equations for the Shear Capacity of Concrete Members without Shear Reinforcement. CEB-Bulletin 224, Lausanne, 1995.
2. MEYER, J. Improving the Ductility of High Performance Concrete (HPC). Vol. 1, Lacer 96.
3. MAYER, L. Neue Entwicklungen beim Einsatz von Hochleistungsbeton. Philipp Holzmann Review, Januar 1996.
4. RAHLWES, K. Recycling von Stahlbeton- und Stahlverbundkonstruktionen. Philipp Holzmann Review, September 1992.
5. KORNADT, O. Advantages and Costs of Sustainable Building Construction. FIP conference in Johannesburg, 1997.
6. KORNADT, O. Nachhaltige Entwicklung: Herausforderung und Chancen für die Baubranche. Bauingenieur to be published.
7. GRÜBL, P. Baustoffkreislauf im Massivbau. Bauingenieur 72, 1997, pp 425-430.
8. BREITENBÜCHER, R. Anwendungsmöglichkeiten für Sifcon. Philipp Holzmann Review, June 1997.
9. KORNADT, O. Gebäude von morgen: Forschungsbericht. Beton-Verlag Düsseldorf, 1997.
10. PHILIPP HOLZMANN AG. Der neue Zollhof in Düsseldorf. Philipp Holzmann Review, Technischer Bericht, April 1999.

PROVIDING CAREER SKILLS AND PROFESSIONALISM IN CONCRETE

J Tolloczko
The Concrete Society
United Kingdom
M Temenides
CIMBETON
Y Malier
Ecole Normale Superieure de Cachan
France

ABSTRACT. Training has always been recognised as an important element in the effective operation of the construction industry. While the need for a well trained workforce, of operatives with good and often specific skills, has been appreciated there have been few incentives to enable, encourage, or force the industry to meet training costs. This has meant that it has not been possible to develop either satisfactory training schemes or systems to provide and administer training qualifications. This situation is changing as client authorities and bodies are beginning to restrict work to organisations prepared to guarantee that the quality and standards of their work will meet requirements. Such guarantees will only be feasible where the quality of workmanship can be assured and maintained by using suitably qualified operatives and supervisors. This paper provides an overview of the current thinking on this issue.

Keywords: Training, Qualifications, Certification, Concrete, Performance, Quality, Durability, Costs.

Professor Jurek Tolloczko is the Chief Executive and Technical Director of the Concrete Society. He has a wide experience in the construction industry particularly in the concrete, offshore and steel sectors. He is a member of various technical committees including industrial training and academic advisory panels.

Michael Téménidès is the Executive Director of CIMBÉTON, the Association of the French cement industry for promotion and technical information. He has previously worked with various companies: a road contractor, a land developer and other sectors of the construction industry such as: steel industry and precast concrete.

Professor Yves Malier is the Director of Ecole Superieure de l'Enseignement Technique and President of AFREM (the French association for research and study of materials). He is also President of l'Ecole Française du béton.

INTRODUCTION

When reinforced concrete first came into use some one hundred years ago it was rapidly taken up as a construction medium. It offered designers a structurally sound material that could be moulded into a wide range of shapes. This flexibility meant that structural form could be taken out of its "straight-jacket". When first introduced the performance expectations and consistency of both concrete and reinforcing steel were low and this was reflected in a conservative approach in design. Basic cements and the resulting concrete displayed strength gain characteristics that, against a background of lower design demands, gave and allowed room for variability in workmanship - there was some 'worker tolerance'.

As materials manufacture and quality improved higher levels of structural performance became available. Designers eagerly took up each increase in performance to refine and 'improve' designs. Using the newer steels and concretes, designers now take designs closer to the limits that can be achieved by the materials. In many cases these developments have not been accompanied by corresponding improvements in workmanship and 'worker tolerance' has been lost [1].

Against this background there have been instances where the durability of reinforced concrete has fallen short of that expected. This is causing concern both within the concrete sector of the construction industry and to its clients. For the concrete industry this situation is made worse by wider market demands that are forcing clients look better returns on investment through cost savings together with major improvements in the performance. They are seeking; more for less, zero defects and performance guarantees.

The concrete industry can meet these challenges but it is recognised that while cost savings and better and more consistent performance can be achieved through the opportunities offered by improved quality materials this will only be possible when designs allow for buildability and ease of construction and high standards of workmanship and supervision are used.

The key to meeting these requirements is through training.

Rather than just saying that designers should bear in mind the needs of the construction process they must be taught about this process. They need to know about each stage from the selection of materials to the completion of each concrete element within a structure. They need to be provided with information on the whole range of concrete materials and techniques that are available and in development. They need to know how materials are worked with and the difficulties that will face the workforce if and when awkward shapes are called for, particularly if access is restricted.

During construction supervisors need to be trained not only in the skills that they are supervising but also in the means by which they can communicate information and instructions to those they are supervising.

For the construction process to be effective operatives must be trained in the skills needed for their work. Experience is important but it should be based on a sound knowledge of methods and techniques to be used for each operation that they may be involved with.

To enable the concrete industry to fully benefit from training and to avoid costly duplication a system of awarding qualifications, or registration of skills, is required. From an operatives point of view such a qualification system will provide a passport to work when he moves from job to job.

In the following sections of this paper further consideration is given to the needs and the way in which provisions for training and suitable qualifications are being developed.

The Need for Career Skills and Professionalism

The provision of career skills and professionalism is of fundamental importance to the sustainability of the construction industry and its associated concrete sector.

Equipping a workforce with the necessary knowledge and the ability to apply knowledge with confidence at all levels of seniority should be standard practice throughout the concrete sector. By doing this the industry would promote and provide training to sufficient people in the skills needed to develop a world class industry. Such a strategy would prevent increasing workloads on fewer unskilled labour, prevent lower quality/productivity and minimise costly re-work and subsequent legal disputes.

Few would disagree with the logic of the above statement but can we honestly state that our industry actually practices what we all know and often preach! Why in this age when clients increasingly demand more for less, zero defects, and guarantees do they also state that the general quality, quantity and mix of skills in our industry remains inadequate to satisfy their stated demands.

One possible reason is that the pressures on time and finance, for those left in the industry after the recent recessionary years, makes it difficult to pursue additional knowledge development. This, supported by fewer and lower calibre younger people entering our industry, should be ringing some major warning bells!

Providing career skills and professionalism is as applicable to the person placing the concrete as it is to the Director of a company. The difference between the two is the type and breadth of knowledge required. The day that each recognises their respective value and contribution will be a major step forward in developing the provision and acceptance of training. The resulting improvements in quality, saving in materials and time and in particular reduced dispute legal fees would provide significant benefits to exceed the initial costs of training.

A BRITISH PERSPECTIVE

What the British Government and Clients are Demanding From the Workforce

The British government is seeking to shift responsibility for construction training from government to industry. Currently, the perception of government is that the quality, quantity and mix of training in the construction industry remains inadequate and that an increase in skills base is necessary to meet demands placed on it.

A Government Construction Task force, led by Sir John Egan, set out a radical new approach to the construction management process [2]. Part of the ethos in 'rethinking construction' was the reduction in defects. This could only be brought about by improvement in the skill base as 'training and quality are inextricably interlinked'. It was the view of the Task Force that '...preference should be given to contractors who demonstrate that they use trained workers.' One way of achieving this is for major clients to insist that workers hold valid registration cards under the Construction Skills Certification Scheme (CSCS)'.

For its part the client is now demanding evidence of competence in the workplace. It believes that improved performance, less remedial work and 'get it right first time' will reduce costs yet improve quality and productivity and remove the 'cowboy' element.

Major companies and organisations are starting to demand CSCS as a pre-condition of winning work. Their subcontractors must have skills certification cards and this requirement is to be built into future tender documents. The concept is analogous to the introduction of ISO 9000, where major companies demanded that 'no quality management scheme, no invitation to tender'.

Registration

Construction Skills Certification Scheme (CSCS)

Registration schemes have been provided for construction occupations for many years. The aim of the CSCS is to extend certification across the industry as a whole. Launched in 1995, the nearly 30 trades covered can expect to have 100,000 workers registered by the end of 1999.

In Britain the issue of training, assessment and certification is presently receiving great attention. This is being driven by clients demanding more for less money, zero defects and guarantees. This is particularly appropriate to the use of concrete since a historic assessment of the concrete industry's "track-record" could be better.

Current trends/developments are tending for construction (including concrete) to train all operatives working on construction sites to be trained to National Vocational Qualification (NVQ) Levels II as a minimum. NVQ's are presently being driven by the Construction Industry Training Board (CITB) UK and are assessment schemes based on the ability of an employee to demonstrate, by training and experience over a period, that a person is competent to work to a certain performance standard.

Competences contained in NVQs are at the core of the scheme. However, for a limited period those who did not possess NVQs were able to become members through industry accreditation of time spent and experience in the industry - sometimes known as Grandfather rights. All cards now have to be renewed after 5 years.

Construction operatives employed in 5 currently unregistered trades including concreting, have until the 15 September 1999 to register. The CSCS application form must be completed and returned within 3 months of this date, together with a copy of the one-day safety course certificate which is a condition of the membership. Operatives registering after this date will need to gain NVQ Level 2.

National Vocational Qualifications (NVQs)

What are NVQs ?

Construction training is driven by the principal aim of attaining an NVQ assessed in accordance with prescribed national standards. An operative needs to possess a prescribed level of generic skill underpinned by a sound knowledge base.

NVQs show that an individual can do a job, unlike other qualifications that only showed that a course had been completed and a written test passed. In this way, NVQs are not part of a training process but they are an assessment of skill levels obtained, hence their relevance to construction operatives.

The operative is responsible for acquiring the skills and knowledge necessary to meet the standards required by the awarding body, which in the case of construction is the CITB/C&G.

The CITB is responsible for designing a suit of NVQs suitable for the construction industry and takes the role of an 'industry led body'. To maintain standards each NVQ design is assessed and approved by NCVQ (now QCA).

NVQs are structured as follows:

Level 1 - basic foundation
Level 2 - skilled tradesman
Level 3 - skilled craftsman/supervisor
Level 4 and 5 - managerial and professional

Each NVQ is made up of a number of units of competence, several of which are mandatory. Each unit is self standing and can be assessed and gained separately. The units are subdivided into elements of competence, and are defined by a range statement (tasks of element), performance criteria (standards required), performance evidence (practical work), knowledge evidence (summary of background knowledge). These elements are assessed as a continuous process over a period in the workplace or on a simulation project where appropriate. Basically, this is through surveillance and oral questioning to prove the competence of the operative in the element. The assessor will also look at historical evidence e.g., certificates, training records, letters of endorsement from employers. This process leads to a 'notification of achievement' and 'certificate of unit credit' which is logged by awarding body who will award NVQ when requisite number of units has been obtained.

NVQs are designed for assessment in the workplace hence training towards each element, unit and NVQ is provided by employers. This requires trainees to be on site with more experienced operatives.

NVQs for Concrete Operatives

A level 2 for Construction Operations has been in-place and available for the past 3 years. Concreting is covered under the General Operations NVQ, which has been popular with the utilities industry and training colleges. This is not considered as providing the practical experience or the necessary skill levels to meet the needs of a vocational qualification.

The NVQ specific to structural concrete 'Concreting' has had a lower uptake although more specific to the concrete industry. Its structure has now been reviewed and updated in light of industry comments and experience and a greater interest in this NVQ is now more likely.

This will be the NVQ 'Structural concreting' that many clients will require concrete operatives to hold.

At Level 3, at present, no specific concrete sector NVQ exists. An NVQ Level 3 in non-specific Specialist Operations has been designed as a framework into which the specific needs of a sector such as the concrete sector of the construction industry can be introduced. This is ideally suited to the specialist. The methodology has already been successfully used by Piling, Tunnelling and Sprayed concrete sectors. It takes the worker beyond the operative role and up to supervisory level including; site establishment; identifying necessary resources; contributing to the work, quality, relationships and information flow.

What Does the Operative Get From an NVQ?

The most obvious benefit to the operative is a registration card or passport to work. It will enhance his status through a qualification that recognises continued improvement and recognition for skills and knowledge gained. Each unit and element is held by the employee and is fully transportable between companies. This is obviously important in an industry that relies on flexibility in the movement of operatives.

Professional Development

In Britain vocational training systems for professionals fall under the general title of 'Continuing Professional Training' (CPD). CPD is the systematic maintenance, improvement and broadening of knowledge and skills, and the development of personal qualities necessary for the execution of professional and technical duties throughout a person's working life. The basis of CPD is that it is never too late to learn or improve, and it applies to all ages and levels of seniority.

CPD may be achieved by attending or undertaking:

- courses, seminars and technical meetings organised by the professionals; institutions/bodies, universities, polytechnics, and professionally recognised course providers.
- conferences, committees and working parties in a managerial capacity;
- private studies of a structured nature;
- correspondence courses, Open University courses and other supervised study packages;
- research and post-qualification studies;
- technical authorship or preparation of lectures for an organised event.

The above relates to all aspects of civil, structural, marine and other forms of engineering. In most cases the use of concrete can be demonstrated. It is therefore essential that the workforce, at all levels of seniority, needs to have the knowledge and skills for the safe, durable and sustainable use of concrete.

The major certification structures for British professionals are defined and administered by:

1. The Institution of Civil Engineers
2. The Institution of Structural Engineers

Each has a scheme that is established and recognised and based on clearly defined criteria relating to:

- A minimum standard of graduate qualification;
- A specific requirement to demonstrate appropriate CPD and that key training objectives are achieved:
- A clear demonstration of experience in both design, construction and management;
- A phased examination and assessment process.

Although not specific to concrete, professionals that are involved in design or construction of concrete structures would have to have the appropriate CPD and experience in concrete. The resulting qualification is membership of a professional institution and registration as a Chartered Engineer.

Concrete Technologists

In Britain the Institute of concrete Technology (ICT) awards membership on the successful completion of Advanced Concrete Technology (ACT) Diploma. This diploma requires candidates to attend approximately six weeks of intensive learning, carry out project work and complete an examination. The ACT Diploma is the only concrete technology qualification available to technical grades but is not mandatory and does not attract the same status as qualifications obtained through the Civil or Structural Institutions.

The Issues Facing the British Concrete Industry

The British concrete industry is facing major changes with clients now demanding:

- more for less money
- zero defects
- guarantees

These demands will result in changes to the concrete industry's operations and there will be a greater emphasis on the use of a better and fully trained workforce. At present there is a shortage of people at all levels (site operative to managerial). This means that the industry must first identify how it can best respond to the increasing demands from clients with the few trained people available.

The main problems are:

- fragmented concrete industry,
- shortage of skilled people,
- lack of resources to respond as a cohesive industry,

- lack of quality teaching at undergraduate level of the “real” issues that face the concrete industry e.g., concrete as a material, design for constructability, quality/durability issues, sustainability,
- lack of industry co-ordination and standardisation,
- lack of time and resources for people to train due to ever increasing workloads.

A FRENCH PERSPECTIVE

Government Policy and Incentives Given to an Employer

Training and education for employees is encouraged by the French government. The companies pay a tax, ‘Taxe d’apprentissage’ to help training centres. Each company has a budget, ‘budget formation’, that benefits from tax incentives where money is used to support training and development.

But these are general regulations. There are no specific settlements for construction. At present students are not being attracted into the construction professions and the lack of direct investment in construction industry training will not help change this trend.

The perception of what is required is different, according to the status and role of those involved in construction:

- owners (investors, users) have strong requirements about: concrete performance, durability, low maintenance, value of the investment, security for users and goods,
- designers also have strong requirements about: aesthetics, shapes, colours and surface aspects,
- Contractors have variable requirements:
    - in civil engineering, for bridges for instance, their requirements are similar to those of designers,
    - for buildings, contractors buy mostly on the basis of price: the consequence is that ready mix industry or precast industry sell them priced goods rather than good quality products.

The Involvement of the Concrete Industry

The involvement of the concrete industry with issues related to the performances or the quality of concrete is variable.

Its involvement in training and instruction is generally low in accordance of the size of the companies.

But the involvement of the cement industry is high:

- Through ATILII and research centres of cement producers, in the development of concrete,

- Through CIMBÉTON and 'l'Ecole Française du béton' (EFB - the French School of Concrete, see below) for construction,
  - Training courses for students in Architecture,
  - The CIMBÉTON contest "Bétons, matière d'architecture",
  - Training courses in high schools for engineers,
  - Training courses for teachers.

The Role of Universities and Schools

Universities and schools summarise the major technological developments and re-inject and reprocess this information for use in basic courses and professional training packages. The EFB ensures that there is educational and engineering and co-ordination.

Relations Between Scientists, Technicians and Professionals of Concrete

These occur through:

- Cement manufacturing groups
- ATILH : Association created by the cement industry for technical research
- National research projects
- Examples: Calibé, BHP 2000, BEFIM (formerly Voics Nouvelles du Béton), FABAC (fatigue tests of continuously reinforced concrete pavements)
- Specific partnerships between firms.

In France, an important contribution to the improvement of relationships was through the establishment of the EFB.

l'Ecole Française du Béton (EFB):

- provides an organisation that encourages thought and activity, and links together education organisations, research laboratories, construction companies, design departments, architectural agencies, and government services......this means that it is able to spread and promote knowledge in the field of concrete construction,

- has created a school with no restrictions, open to all; architects, engineers, technicians, teachers and researchers,

- provides a space for initiatives, so that anyone may initiate vocational training at a local, regional or national level, and help build up a network between schools, universities and technical centres.....this enables the identification and optimisation of synergies and the best possible solutions for training needs to be found,

- acts as a tool for dialogue and a catalyst in helping schools, universities, scientific associations, trade organisations to improve their operations and develop their capacity to lead in the development of vocational training that will appeal to all construction project participants.

The Further Aims of the EFB are to:

- inform education organisations involved of innovations in the field of concrete structures and materials,
- assert French expertise in the field of concrete construction, by spreading knowledge in France and abroad,
- reinforce communication between its members and partners

Through the EFB :

- architects, engineers and technicians have access to up-to-date information on recent projects and on the applications of research findings,

- teachers have access to design aids for their courses and practical works, these aids taking account of statutory, technical and scientific evolutions,

- owners benefit from decision support through provision of and access to relevant information,

- companies benefit from the training of their employees,

- schools and universities can use the results stemming from research, especially on national projects, for their professional courses,

- the French construction industry can give a more modern and dynamic impression of itself and its activities.

Various EFB Actions Include:

Organisation of seminars, to spread and promote the results of national projects, amongst professionals within the construction business.

University winter and summer courses,

Support modules for courses on concrete,

Development of educational tools and materials,

Promotion of educational and reference books labelled by the EFB.

Examples of EFB Activities in 1998

- Creation of a news bulletin, a link between engineers, researchers and teachers.
- Regional seminars for technical teachers: (more than 500 participants*).
- Regional seminars for engineers from the Direction Départementale de l'Équipement and territorial organisations (more than 400 participants*).

- A winter and a summer school for the vocational training of technicians, where practical projects were examined and discussed including: "transport and pouring" and "durability" of concrete.
- Launch of a collection of reference books, written in collaboration with several specialists, and labelled by the EFB.
- Information seminars for teachers on the use of Eurocodes.

*Each participant has been provided with documents edited by CIMBÉTON, especially practical manuals for the design and execution of works.

APPROACH TO TRAINING IN SOME OTHER EUROPEAN COUNTRIES

The need for training provision is a challenge that does not recognise national boundaries [3]. The problems are common to many countries. In some instances there are also moves towards a more positive approach to training provision with some signs of direct support from industry and government.

In countries such as Holland and Germany there are strong links between government authorities and the National Concrete Societies and the Societies play a direct role, particularly in the case of Germany, in the development, monitoring and control of training and qualification.

Holland

The Present Vocational Training System in Holland

The vocational training system is based on a system of qualifications. Training at all levels can be offered by all kinds of organisations, varying from universities and technical schools to branch - organisations and institutes. However, all have to meet requirements for final attainment-levels that have been or are being developed. These attainment-levels are checked and have to be recognised by the 'Landelijke Organen' (a National Government Organisation).

The qualification-structure that is being used in Holland is an adaption of the SEDOC-system of European Qualification Structures. Within SEDOC there are five levels of education and training. Within this approach it is not the education/training programmes that are assessed and ranked, but the way in which these will lead to a qualification that measures up to the specified final attainment level requirements.

The list of requirements for the attainment levels are based on job profiles and the necessary skills to perform the tasks to meet a job's needs. Job profiles are being developed through a series of job analyses. The job analyses for the concrete sector is being performed by Betonvereniging - the Dutch Concrete Society. Basic elements from this work will be used within a EC funded Leonardo Project on Advanced design and construction in concrete to provide information job analyses and to help identify training needs within Europe.

In addition to the development of a wider range of vocational training there are already a number of organisations that offer post-university courses on concrete design. Information on the numbers attending, course contents, quality, examinations etc., has to be given to a special government commission that checks to see that required standards are being met.

The Problem Facing the Labour Market in the Field of Concrete

The main problem facing the industry is a lack of skilled people at all levels both now and in the near future. In the industry the organisation of design and building processes is changing. In Holland new and different forms of contracts are being introduced, so contractors now need people with higher levels of organisational skills. This is reflected in the design of courses in higher and a tendency for graduates to be more skilled organisationally but less well trained technically. However, the biggest problem is the shortage of people coming into the industry from school and university. The image of a career in the construction industry is poor and this is an issue that must be addressed particularly as economic growth means that in the future there will be a need for more not less people.

Training in Concrete

At post-graduate level further education is required for the development of specific and more purely technical skills.

From the industry's point of view still the training of new employees within a company takes a lot of time and effort (1 to 2 years). There needs to be more incentives for companies to invest in training. These incentives may be developed by client pressures particularly from client authorities.

In the future training will be required in the environmental aspects of the industry's activities. This will flow from the need to meet requirements in the field of sustainability. Building materials, and working conditions. Requirements in these areas will be imposed within a few years through such agents as Eurocodes. In Holland the industry is preparing itself to fight off what might be unwanted and too complex legislation in these areas.

Betonvereniging's Involvement in Training.

The Dutch Concrete Society is involved in the development and staging of about 42 courses at various vocational levels. It is the main player in the field of vocational training on concrete. It is involved in the regular education system by the Government through its delegates in Commissions, etc. It is the organisation within the concrete industry that performs job analysis and co-ordinates the composition of final attainment levels that are used to in the development of concrete courses.

The Society is involved in higher, post-university, education through its close co-operation with universities and other national training and educational bodies. In particular it is the driving force behind the development of the post-graduate course on structural concrete design leading to the qualification of 'Professional Master of Structural Engineering.

Monitoring Training Needs

The Society sets up a special expert - commission for each course that meets twice a year to examine the continuing relevance of course content and if changes should be made. The commission does this by drawing on information from industry through the Society's member companies network and by working closely with branch - organisations and contact with bodies in the regular education system.

Trends and developments that will influence future education and training within the concrete industry.

- More international co-operation and competition: there will be a need to have more knowledge of languages and foreign legislation.
- Companies will get bigger: there will be a trend towards fewer but bigger companies that will offer all services, and specialised smaller enterprises.
- Eurocodes: when they are brought into use there will be a strong demand for additional training.
- Automation and IT involvement: this will increase and all operatives will need appropriate training.
- Material properties will continue to be developed: training must be available to keep the industry up-to-date with these developments and the subsequent developments of new building methods and techniques.
- Environmental and performance related issues: to meet changing attitudes towards the environment and the use of natural resources training must be developed to keep the industry well informed and up-to-date on such issues.
- More attention will be focused on safety issues: there must be a continuing emphasis on the provision of training for safety to reduce injury levels within the construction industry itself as well as within the concrete sector.

Germany

Training and Qualifications in Germany

Post-university courses on concrete design and the internal qualifying system of the contractors are under the responsibility of the German Concrete Society - Deutscher Beton-Verein EV (DBV).

To understand the vocational training system for concrete design it is necessary to recognise the different requirements and factors that influence what needs to be done and how it is done.

- *Legislation.*
  Various construction regulations not only control technical matters, they also call for operational qualifications. They contain specified levels of qualification required by producer, contractor or operators for different areas of work in the various fields of concrete construction.

- *Codes, guidelines, technical contract conditions.*
  There is also a wide range of construction directives that give details about the requirements for the qualification and training of those people or contractors, involved in the design and construction of concrete and for those involved in the repair and maintenance of concrete structures.

Control of Training and Qualifications

To meet the needs for suitable qualifications within the industry the DBV has two main Advisory Committees that set, monitors and control standards for the quantity and quality of required knowledge and experience.

Since 1968, the professional qualifications and experience of the concrete engineer has been established by the DBV's Advisory Committee on Concrete Technology - 'Ausbildungsbeirat Beton beim DBV'.

Since 1985, the professional qualifications and experience of repair and maintenance specialists has been established by the DBV's Advisory Committee on using plastics in concrete construction - 'Ausbildungsbeirat Verarbeiten von Kunststoffen im Betonbau beim DBV'.

The DBV's Advisory Committees are made up with representatives of several industry associations, appropriate ministries, building authorities and universities.

The DBV, working with other industry based organisations also shares responsibility for qualification of personnel and contractors working in such fields as new concrete construction in contact with water or the execution of repair and maintenance works on water retaining structures.

Certification Structure / Requirements and DBV's Involvement in Training

The DBV is involved with many training activities within the industry and within the main stream educational establishments in Germany.

Its own activities in the field of training and qualifications include:

Certificate for extended knowledge and experience on concrete technology - 'Bescheinigung iiber den Nachweis erweiterter betontechnologischer Kenntnisse'.

Certificate for protection, repairing, connecting and strengthening of concrete structural elements - 'Befähigungsnachweis zum Schützen, Instandsetzen, Verbinden und Verstärken von Betonbauteilen'.

Certificate for shotcrete works with plastic modified shotcrete - 'Düsenführerschein'.

The training centres of the construction industry stage courses for these qualifications based on the DBV's education and examination regulations. All certificates are registered by the DBV.

Every year the DBV organises workshops, regional and national conferences to educate, inform and provide guidance to the personnel of consultant and contracting organisations.

The DBV's advisory engineers also carry out internal training courses for its members.

The Problem Facing the Labour Market in the Field of Concrete

In spite of the drive to ensure that only suitably qualified people and organisations are used for construction projects there is still a problem where unqualified and inexperienced competition put up low price bids. This can mean that qualified organisations using qualified personnel are priced out of the market. To meet this difficulty more and more contractors are tempted to work with unqualified or poorly qualified staff to keep their prices down. This has negative effects on the quality of concrete structures.

The Main Problem in Concrete Education?

There are some fears in Germany that education in the field of concrete technology is being reduced in favour of design and construction.

Developments in Legislation / Requirements that Will Impact on the Concrete Industry

The new directives (codes, regulations) as a result of the European Construction Product Directive will have a direct impact on the concrete construction market. In addition to this new laws on conservation, new health and safety regulations will have a direct influence.

Publications

The DBV publishes numerous books and guides including a technical leaflet series.

Monitoring Training Needs

The DBV's Advisory Committees monitor training courses and up-date the education plans in response to industry requirements.

The DBV's Advisory Engineers analyse the effectiveness and look for defects in concrete education based on opinion polls taken at the end of regional conferences.

THE WAY FORWARD

The need for training is generally accepted and is becoming established throughout Europe. There are still difficulties in finding the means by which those who invest in training to improve the performance of the construction industry receive a benefit rather than a competitive penalty for their efforts. Training needs are being identified and it is known that many of these needs can be fulfilled.

Suitable vocational qualifications are being developed and systems for the registration of skilled operatives can easily be brought into place. When government, client authorities in both the private and public sectors, other sector clients and the construction industry are ready to face up to the cost involved in the knowledge that there will be a long term benefit in improved performance - then training will get the priority it deserves. When this happens the concrete industry is ready to respond and well placed provide the required training.

REFERENCES

1. WALKER, M J. Concrete Quality and Quality Enhancement - Sound Concrete, Durability Issues and Strategies for Good Performance. Paper Presented at the NRMCAM 1998 International Concrete Convention, Kuala Lumpur, Malaysia, 8th & 9th October 1998.

2. EGAN J. (Chairman). Rethinking Construction - The Report of the Construction Task Force to the Deputy Prime Minister on the Scope for Improving the Quality and Efficiency of UK Construction, London, The Stationery Office, 1998.

3. Responses to Questionnaire about Training Needs and Provision in European Countries. Unpublished Papers of the Committee for the EC Funded Leonardo Project on Advanced Design and Construction in Concrete.

LEADER PAPERS

RADICAL DESIGN AND CONCRETE PRACTICE

R S Narayanan

S B Tietz and Partners

United Kingdom

ABSTRACT. Designing for sustainability will force all industries to re-think their current methods of design and construction. Life cycle analysis will be necessary to identify and minimise environmental impacts. Recycling materials and sites will also become the norm in the future. Designers can contribute by re-examining the design loadings and partial factors. The industry is taking steps to improve the quality of concrete construction. This in turn will also contribute to sustainability.

Keywords: Sustainable construction, Environmental impact, Life cycle analysis, Re-cycling, Design loads, Partial factors, High strength quality control issues, Whole life Costs, Certification, Operative training.

Professor R S "Nary" Narayanan is a Partner of S B Tietz & Partners, Consulting Engineers, London. He is a Visiting Professor at The University of Leeds in Principles of Design. He sits on a number of Technical Committees and chairs some of them. He is currently Chairman of The Concrete Industry Alliance Steering Group for Concrete 2005, the Innovation and Research Strategy.

INTRODUCTION

Concrete is an exceptionally versatile construction medium. It has found application in every major form of construction. It combines strength with durability. It can be moulded into shapes to produce aesthetically attractive structures. Many elegant, delicate and monumental concrete structures all around the world bear testament to the qualities of concrete. Concrete can be made with different properties to suit particular applications (e g) densities, strengths, pore structures. It can be pre-cast or placed in-situ. It can be pumped sprayed and placed under water.

In its long history, concrete industry has not been without its problems. This is not uncommon with any technological development. The press, with its love of "bad news", has helped to create a poor image for concrete from time to time. The media are not interested to note that the majority of concrete construction performs well. Responsible sections of the industry address the problems that manifest and analyse their causes. Where new measures are warranted, these are put into practice. It is certain that the industry will face challenges in the future and it will need to be constantly vigilant and be prepared to adapt.

A challenge, which is at our doorstep now, arises from society's awareness and concerns regarding environmental matters. A welter of legislation already exists in this area. I am sure that more is in the pipeline. The UK government is a signatory to a number of international treaties in this field and has committed itself to a number of environmental targets. Sustainable development is now more than just a slogan. This has enormous implications to the construction industry.

All sectors, be they material producers, designers, contractors or client bodies, will need to take into account the requirements of this extra dimension of sustainability in their sphere of activity. A number of studies and research are already underway to understand the issues for concrete construction and to provide methodologies to facilitate decision making.

In this keynote paper, I propose to survey some of the initiatives the industry has undertaken with the hope of informing a wider audience

PROFILE OF THE INDUSTRY

In any broad discussions covering the whole of an industry, it will be useful to have a feel for some vital statistics. I have drawn heavily from Concrete 2005 (DOE's Innovation and Research Strategy for the Concrete Industry Sector [1]) with which I was closely connected. Figure 1 shows the complex web of sectors which make up the industry. Figure 2 shows an overview of the industry, represented by a cube, the three axes of which being markets, materials and processes. Decisions are made at each of the blocks, which affect the use and performance of concrete in respect of other blocks. Figure 3 provides a feel for the amount of materials used in concrete construction. Figure 4 is a pie chart of the industry by types of construction

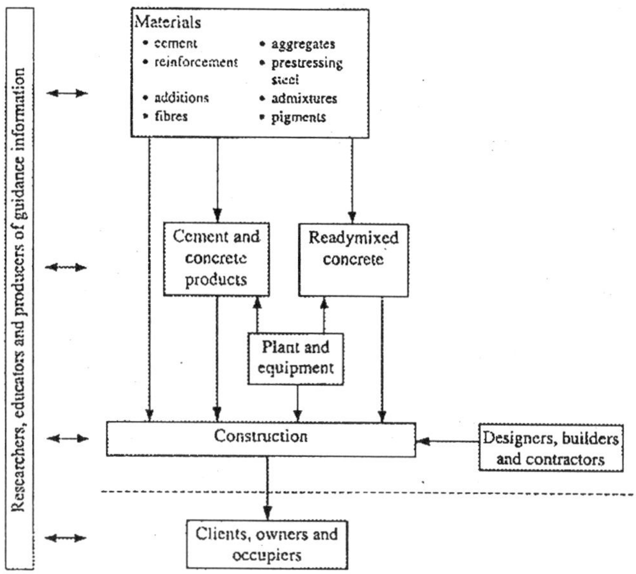

Figure 1 Definition of concrete industry sectors

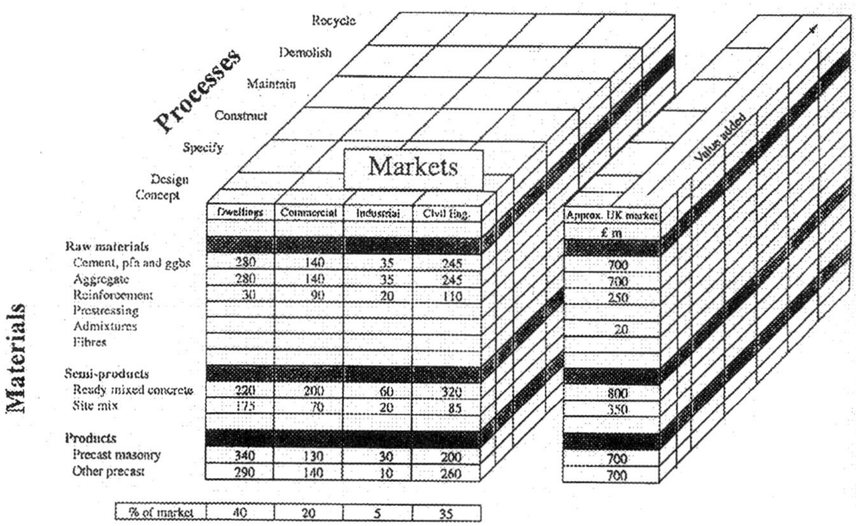

Figure 2 Markets for concrete industry - 1994

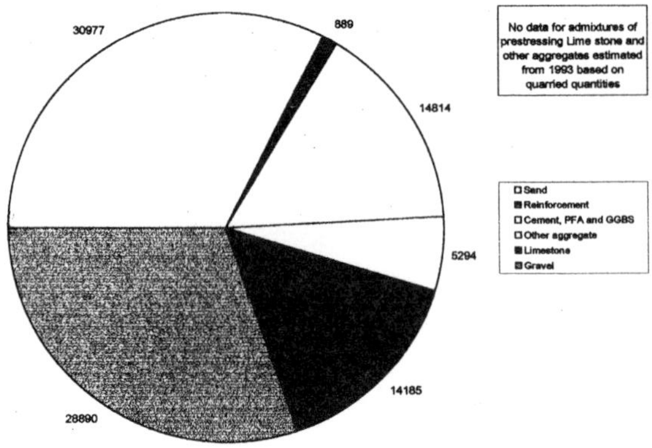

Figure 3 Material used in concrete construction, 1994 (Units in 1000's tonnes)

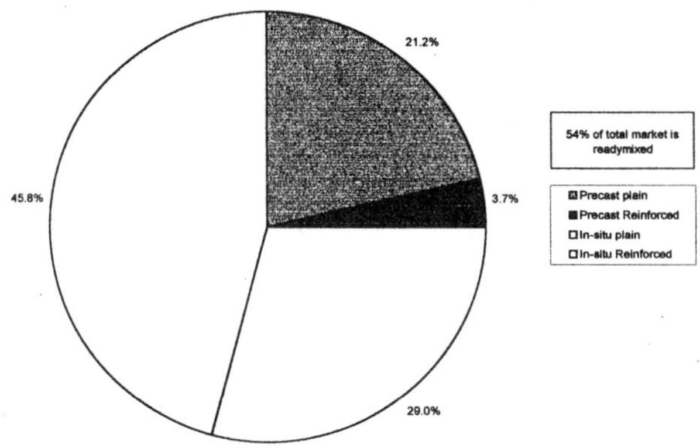

Figure 4 Ratio of types of concrete construction, 1994

SUSTAINABLE CONSTRUCTION

The concept of sustainable construction embraces a number of objectives [2]. Firstly it aims at social progress, recognising the needs of everyone. The objective simply means that everyone should be able to live in a clean and safe environment. The second strand is the effective protection of the environment from global and local threats. This includes the protection of things we value such as wildlife, landscapes and historic monuments. The third aspect is the prudent and efficient use of natural resources. This in turn requires the development of renewables as alternatives to materials and fuels of limited supply. Sustainable development also aims at the maintenance of high and stable levels of economic growth and employment.

These new imperatives have enormous implications for the construction industry but they do not imply that we abandon the way we live now and resort to "cave-man" technology. Rather they demand that we become more efficient and competitive through innovation and better management of our affairs. Everyone in the supply chain of the industry (Figure 1) has a role to play. Considerable research is already underway to facilitate sound decision making. As with all innovations, risks will need to be balanced against potential gains. However, the alternative of doing nothing is not risk-free. It is simply not sustainable. Let us now look at what is possible.

Environmental Considerations

A few bold statistics will show: (a) how environmental considerations are relevant to construction activities, and (b) the scope to contribute to the well-being of the society.

- Energy use in buildings account for between 40% and 50% of the UK emissions of carbon dioxide (CO_2)

- Construction industry consumes around 6 tonnes of material per person per year. This simply dwarfs the amount of material used in any other industry

- About 70 million tonnes construction waste is produced every year in the UK

Thus the scope for good housekeeping, reducing consumption, reducing waste and re-cycling is considerable.

Incorporation of a formal environmental component into construction design will become increasingly important and necessary. Before looking at the steps involved in this process, it will be useful to have an overview of some selected environmental indicators and constructionrelated contributions. This is shown in Table 1 [3].

It will be seen from the table that:

- Energy associated with the provision and use of buildings and structures accounts for about 77% of UK final energy
- Energy associated with construction materials account for about 7.9% of the construction energy (provision and operation)
- Operational energy in running the buildings far outweigh the embodied energy (de) the energy consumption in the production and transport of construction products and material

Life Cycle Analysis

One method of analysing and assessing the environmental impact of a product or service is Life Cycle Analysis (LCA). The International Standards Organisation (ISO) and the Society of Environmental Toxicology (SETAC) are developing guidelines.

Table 1 Seclected environmental indicators and construction-related contributions

INDICATOR	UK TOTAL	CONSTRUC MATERIALS	CONSTRUC PROCESS	BUILDINGS IN USE	DWELLINGS IN USE	TRANSPORT	OTHER SOURCES
Final energy PJ (%)	6283 (100)	386 (6.1)	54 (0.9)	952 (15.2)	1839 (29.7)	1676 (26.7)	1376 (21.9)
Carbon dioxide, Mt (%)	548 (100)	37 (6.8)	4 (0.8)	77 (14.1)	149 (27.1)	145 (26.4)	136 (24.8)
Sulphur dioxide, kt (%)	2718 (100)	151 (5.6)	21 (0.8)	372 (13.7)	719 (26.4)	251 (9.2)	1204 (44.3)
Nitrogen oxides, kt (%)	2218 (100)	231 (10.4)	8 (0.4)	134 (6.0)	259 (11.7)	1356 (61.1)	230 (10.4)
VOCs, kt (1%)	2220 (100)	94 (4.2)	36 (1.6)	27 (1.2)	53 (2.4)	938 (42.3)	1072 (48.3)
Carbon monoxide, kt (%)	4833 (100)	629 (13.0)	3 (0.1)	54 (1.1)	281 (5.8)	3786 (78.3)	80 (1.7)
PM10, kt (%)	250.0 (100)	26.1 (10.4)	4.1 (1.6)	14.5 (5.8)	28.0 (11.2)	71.0 (28.4)	106.4 (42.6)
Heavy metals, kt (%)	4.099 (100)	0.689 (16.8)	0.007 (0.2)	0.120 (2.9)	0.231 (5.6)	0.245 (6.0)	2.808 (68.5)

There are four steps:

1. Definition of goal and scoping: This entails specifying a functional unit (viz) a product or servlce for analysis (e g) supply of one cubic metre of concrete with a specified strength.

2. Inventory analysis: This comprises the quantification of the inputs to and outputs from the system required to supply the functional unit. The inputs will include materials and energy and outputs will include waste and emissions to air and water. See Figure 5 which illustrates the processes involved in the supply of ready-mixed concrete.

3. Impact assessment: Here examining the inventory differences and various environmental indicators on a weighted basis (greenhouse gases, acidification etc) compares alternative functional units.

4. Improvement assessment: Here there are opportunities to improve the cradle-to-grave environment performance.

Some software is commercially available, mainly outside the UK. Currently a multipartner project, sponsored by DETR, is underway. It is in its initial stages. It hopes to define environmental performance in the concrete industry in the UK. It will assess the environmental impact of a variety of concretes and concrete elements. A typical output is shown in Figure 6 [4].

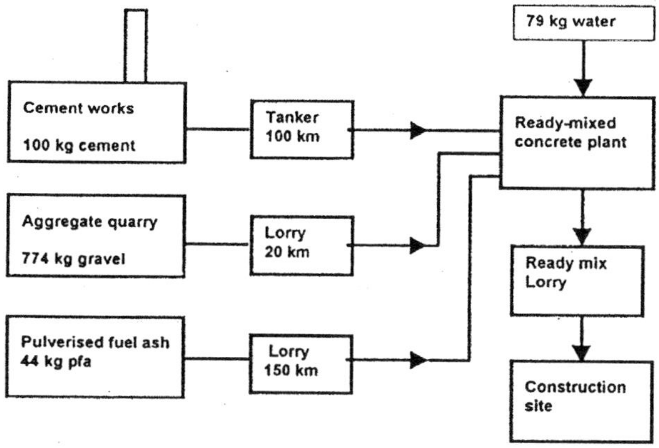

Figure 5 Process diagram for the supply of 1 Tonne of ready-mixed concrete

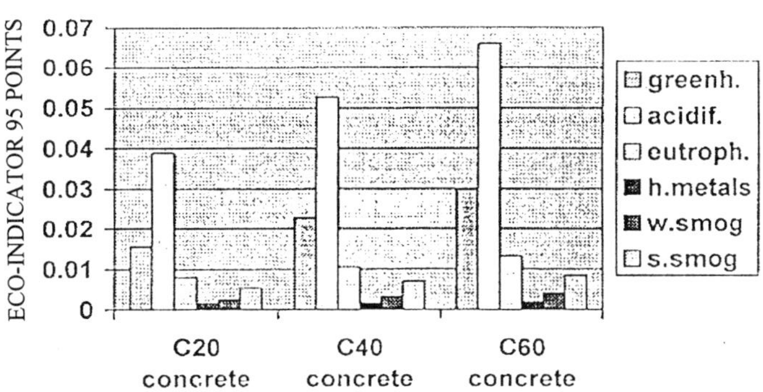

Figure 6 Evaluation of concretes based upon the ECO indicator 95 method for different indicators

Although Figure 6 suggests that stronger concretes are less eco-friendly, the picture looks very different when durability and service life are taken into account and usage of materials on a "per year" basis is considered. There is a non-linear relationship between service life and the usage of materials. As an example, Parrot has considered the durability and environmental profiles of reinforced concrete columns (Reference 5). Service life was assumed to be limited by carbonation induced corrosion of reinforcement and the formation of visible cracks.

Parrot found that only small increases in concrete quality were required to increase greatly the service life. He evaluated the concrete quality and reinforcement required for different periods of service life for a column of a given height to support a given load. In turn he computed the cost and environmental impacts. The results are shown in Table 2.

Table 2 Material costs and environmental impacts per year of service life

SERVICE LIFE (YEARS)	28 DAY CUBE STRENGTH (N/mm²)	AMOUNT FOR 1 COLUMN PER YEAR OF SERVICE LIFE		
		COST (PENCE)	ACIDIFCATION IMPACT (kg SO_4)	GREENHOUSE GASES (kg CO_2)
100	39.0 [CEM 1]	55	0.022	2.7
80	38.2 [CEM 1]	69	0.027	3.3
60	37.0 [CEM 1]	94	0.036	4.4
40	35.3 [CEM 1]	146	0.055	6.6
20	32.0 [CEM 1]	299	0.111	13.1
100	47.0 [CEM 2]	48	0.018	2.2
80	45.9 [CEM 2]	61	0.022	2.7
60	44.2 [CEM 2]	84	0.031	3.6
40	41.8 [CEM 2]	128	0.045	5.2
20	37.9 [CEM 2]	272	0.092	10.4

Notes: CEM 1 - Portland Cement
CEM 2 - 30% of the Portland Cement replaced with PFA.

This is a clear demonstration that: (a) higher strengths confer longer service life; (b) cement replacements have beneficial effects on cost and environment; and (c) environmental assessment should be studied in conjunction with service life.

Recycling

One of the main strands of sustainable construction is minimisation of waste, which in part can be achieved through recycling. The waste thus becomes a resource, enabling reduction in demands on natural resources.

Concrete industry already uses recycled material. Few people know that the bulk of reinforcement is produced using scrap metal as the raw material. Recycled concrete aggregates (RCA) are used mainly in road construction and in some applications of plain concrete. More should be possible, starting with situations where performance requirements are modest. To this end DETR has commissioned research in Dundee University. Their initial findings are that: (a) the density of RCA is marginally less than that of natural aggregate (NA); (b) water absorption of RCA was found to be six times higher; (c) RCA concrete specimens were more porous compared with the equivalent concrete using natural aggregate NA; (d) with a mixture of RCA and NA workability was found to reduce with increasing proportions of RCA. Beyond about 30% RCA concrete was found to be "harsh" (less cohesive) leading to higher "bleeding" (loss of water); (e) Compressive strength of concrete with up to 30% coarse and 20% fine RCA was comparable with NA concrete. With higher proportions a reduction in strength of about 10%-15% was observed; (f) Regardless of RCA source, the variability in composite strength of concrete with a range of RCA content was found to be negligible, when compared with corresponding NA concrete.

Some further tests are planned. The specifiers should be receptive to the information and carefully study them. Then they should adopt them for appropriate applications. Initially there will be some inhibitions. Perhaps DETR might consider a demonstration project to improve confidence further. For instance, a structure using recycled materials can be built at Cardington.

Recycling pre-cast building components should certainly be possible. After all, salvaged bricks are re-used. Components will need to be designed and detailed with recycling in mind. Minor modifications of ends of units may be required. However, proper labelling and maintenance of full records will be essential. Other procedures will evolve when the society realises the feasibility of recycling. What is now expensive, will become economically viable as alternatives become relatively more expensive or prohibited by law.

In a similar vein, sites which were considered unviable not so long ago because of contamination, are now being developed more routinely. The government has signalled through their consultative paper "Planning for the communities of the future" its desire to achieve 60% of all new housing on brownfield sites.

Use of cement in this area, to clean and reclaim contaminated sites, has not been fully exploited in the UK, compared to some other countries, such as the USA. The conventional technique of"dig and dump" simply transfers the problem elsewhere. This is environmentally unfriendly and cannot be sustained. There are a number of methods including the following:

- In situ stabilisation and solidification by forming columns of treated soil matrix, using soil mixing augers and injecting cement based grouts.

- The contaminated soil may be excavated, thoroughly mixed with cement and water and the treated soil then placed in layers for compaction in the usual way. Alternatively, the contaminated soil may be excavated and then placed in layers with beds of cement between them and compacted.

- Sludges and sediments in lagoons may be treated directly by mixing with cement slurry which transforms the sediments physically and chemically, resulting in a material suitable for structural fill.

In all these methods, cement based grouts chemically stabilise and physically immobilise the contaminants within a solid matrix. This improves the engineering properties of the ground, minimises leaching and reduces the permeability. Cement based treatment is suitable for a significant number of organic and inorganic containments, particularly heavy metals.

The British Cement Association has taken the initiative to publicise the techniques (6). Designers should consider these alternatives critically and adopt them.

DESIGN

A number of features of the current design practice could be re-examined and more radical approaches could be considered.

Loadings

In commercial developments the client usually stipulates the design superimposed loading to be used. Often these are considerably in excess of the loadings prescribed in BS Loading Codes. The usual reason given is that there may be changes of use in the future which may demand increased loadings. Can we afford this luxury any longer? Will it be more sensible to design the vertical elements and their foundations for some overload but accept that floors may need some strengthening should the need ever arise. Clearly this is a matter for dialogue between the client and the designers. In my experience very few buildings have used the passive overload capacity, in practice. In the context of sustainability such over consumption of materials could be considered wasteful.

The other aspect of design loadings is the application of a uniform load factor (γ_f) for all dead loads, including self-weight of members. In part the load factor caters for variability and uncertainties of loads. Tolerances on the dimensions of members are fairly small and the density of concrete as produced has been fairly consistent. Under the circumstances the case for a reduction of YF is compelling. This is under serious consideration in Eurocodes. In any event, this could be made conditional on verification of construction.

Beeby and Fathilbitaraf [7] have shown that there is scope for reducing the reinforcement in continuous members in bending. This is possible because of (a) redundancy in continuous construction leads to a considerably higher safety level compared to determinate structures and (b) the pattern loadings prescribed in codes are unrealistic. However, if fully exploited this will increase the service stress in reinforcement. Thus serviceability limit state may well control the design. In general this is unlikely to be a problem in beams. Slabs, which account for the bulk of material usage in a building, may in any case benefit from more rigorous serviceability limit state checks for cracking and deflections. This is even more true when realistic service loads are used as in Eurocodes. Currently full characteristic loads (Q_k) are used for serviceability verifications in the UK. Eurcodes use quasi permanent loads (ie) $\psi_2.Q_k$ for reinforced members, where ψ_2 is a factor (less than unity) the value of which depends on the use of the building.

High Strength Concrete

Properties of high strength concrete are now well established through considerable research. On this basis it has been decided to extend the concrete strength class covered in EC2 to C105 (cylinder strength). Many ready mix suppliers are also now familiar with the technology. It is obvious that the higher strength will help reduce the size of compression members. There are many other benefits which are not immediately obvious. High early strength will permit early striking of formwork and speed up in-situ construction. High strength concrete invariably uses superplasticicers and high slump. The increased workability will result in well compacted concrete which in turn will enhance the durability.

In flexural members the increased tensile strength of concrete will tend to reduce deformations. The increased tensile strength could also be exploited in the design of pre-stressed members in some circumstances.

Housing and Concrete

Currently in the UK, low rise housing generally uses cavity walls for the external envelope, timber floors and stud internal walls. Society still generally prefers facing bricks externally. For thermal insulation the inner leaf is generally in lightweight blockwork such as those made out of autoclaved aerated blockwork (AAC). In the Continent the use of concrete in housing is common. Thermal insulation requirements in some of these countries are even more stringent than in the UK. There will be considerable advantage in using reinforced AAC (RAAC) panels for inner leaf and flooring. In this form strength and insulation characteristics are combined. Being pre-cast and lightweight, the structure can be put up with considerable speed. Acoustic insulation use will also be better as the density of floors will be greater than traditional timber construction. It goes without saying that all the components should be properly designed and detailed. This is an area well worth exploring.

Codes

As engineer's work is concerned with public safety, it will always be subject to regulatory control. There are a number of reasons why codes evolved [8]. They include the following: (a) codes enable consistent enforcement of regulatory requirements both geographically and with respect to time; (b) codes can form the basis for tenders and appointments and (c) they provide an agreed basis for design aids and manuals.

As with all standardisation, the question is how far does one take codification? The answer depends upon the quality of the users. If engineers are well educated and highly trained then the code could be a simple set of Principles. In this scenario, every engineer can come up with his own ingenious method for compliance. This is if course pie in the sky! The credentials of the average engineer forces one to conclude that the code should go a bit further than just principles. In fact Eurocodes do just this. They contain Principles with which all designs must comply and also Application Rules, for which alternatives are permitted. Thus Eurocodes are the nearest we are likely to have by way of innovative codes for quite a long time. They are flexible in that they do not inhibit innovation but also provide well proven methods for the routine jobs which are in truth the main stay of the industry.

Specifications

Every designer prides himself with a bespoke specification for every project. When one analyses projects, it is obvious that most processes adopted by the different contractors, irrespective of the origin of the specification, are similar. Standardisation in this area clearly will improve the efficiency of the industry. Like Codes of Practice, it will facilitate clear communication between parties. The recently published "National Concrete Frame Specification" [9], points the way. Part 1 of this document contains all the usual clauses, (eg) materials, workmanship, etc.

It is followed by Part 2, which deals with projectspecific requirements. It lists the information to be supplied to the contractor together with the information to be provided by the contractor. The designer can insert special requirements if any. This seems an excellent start and the industry should discuss where similar approaches may be possible.

QUALITY ISSUES

There is a growing awareness in the industry for the need to achieve good quality construction and getting it right the first time. I will briefly mention three areas which are receiving attention.

Whole Life Costing

Most clients understand that the lowest first cost does not necessarily give the best value for money. Lowest first cost has an appeal when the responsibility for capital and maintenance is not in the same hands (e.g.) different budgets or buildings developed and sold on. Thus a more holistic view is not a consideration. The culture is now changing (slowly) and whole life costs are now considered in some projects. Whereas whole life costing had been used frequently to assess the merits of different design solutions or alternative materials, until recently it had not been fully utilised for assisting investment decisions. It is partly for the reasons cited above and also the methodology had not been fully evolved. Now BCA have prepared a useful report for DETR [10] which links quality to whole life costing. It defines a model for service life at the end of which some intervention is likely to be necessary and provides a method for calculating costs on a normalised basis. When predicting the future there are clearly imponderables but their sensitivities can be tested The industry thus has the tools necessary to take the method further. For bridges, similar methodologies have existed for some time but there the builder and operator are usually the same body. Whole life coatings will be very relevant to PFI schemes.

Certification

There is a desire in some quarters to develop a certification scheme for concrete structures. There is indeed currently a project sponsored by DETR to explore the mechanics of how this might operate. It can only be a good thing. It is possible that only limited operations in the construction process might be covered by this scheme. It will still be better than nothing. After all we have similar schemes for things such as steel reinforcement, ready-mixed concrete etc. When designers and clients demand these, the industry will deliver them.

Operative Training

It is absurd that at this juncture in the history of concrete we still do not have a recognised and proud tradesman called "The Concretor". This is largely the fault of the designers and clients for not insisting on the work being carried out by qualified tradesman.

However, the trend is now changing. Major clients are now demanding this and the contractors are beginning to respond. There are also moves afoot to support and enhance some of the NVQ's to facilitate the culture changes needed to achieve quality.

CONCLUDING REMARKS

The survival of the industry demands innovation and radical thinking. I have touched upon a number of current issues exercising the concrete industry. Sustainable construction will force us to consider the environmental impact, use of recycled materials, and minimisation of waste. Design and quality issues will also contribute to sustainability. In design use of more realistic loads and load factors should lead to economies and contribute to sustainability. Potential use of concrete in housing needs serious consideration. Specifications could be standardised to achieve clear communication. Quality of concrete construction will be improved when the significance of whole life coatings is fully appreciated. Certification schemes for contractors and the use of qualified operatives will further underpin quality.

REFERENCES

1. DEPARTMENT OF THE ENVIRONMENT Concrete 2005: DOE's Innovation and Research Strategy for the Concrete Industry Sector - June 1996

2. DEPARTMENT OF THE ENVIRONMENT, TRANSPORT AND THE REGIONS. Opportunities for change. Consultation Paper on a UK Strategy for Sustainable Construction – DETR

3. PARROTT L, BCA: Environmental Perspective on UK Construction Materials: (CIB World Building Congress 1998, Gavle, Sweden)

4. PARROTT, L. BCA: Environmental Performance of Concrete - A Partners in Technology Project

5. PARROTT, L. BCA: Durability and environmental profiles of reinforced concrete columns, 6th LCA Case Studies Symposium SETAC-Europe 1998

6. BRITISH CEMENT ASSOCIATION. Recycling Contaminated Land using Cement

7. BEEBY, A W, FATHILBITARA, F "The Design of framed structures - A proposal for Change" Int Conference. "Concrete in the Service of Mankind" University of Dundee, June 1996 pp 495-506.

8. NARAYANAN, R S: "Structural Codes of Practice - A UK point of view" - Structural Engineers World Congress, July 1998 p 318-2.

9. National concrete frame specification for building construction - Construction Research Communication, 1998.

10. BRITISH CEMENT ASSOCIATION. Relating quality to whole life costing - Report to DETR - July 1998.

RESEARCH DIRECTIONS FOR LIFETIME PREDICTION AND LONG-TERM DURABILITY

T Igusa

S P Shah

Northwestern University Illinois

United States of America

ABSTRACT. There has been increasing concern on the long-term performance of concrete and other cementitious materials. Herein, a few possible research directions are summarized with applications to chloride permeability, freeze-thaw and corrosion of reinforcement. The emphasis is on combining non-destructive and other testing techniques with statistics-based reliability methods. It is believed that lifetime prediction methods should be based on an approach combining materials science and statistical theory. Furthermore, it is believed that novel methods of analysis, using indirect or implicit investigations, are needed for materials as complex as concrete. A brief example of such an indirect analysis is presented.

Keywords: Accelerated testing, Durability, Experimental design, Long-term durability, Non-destructive evaluation, Reliability engineering, Service life prediction, Statistics.

Professor T Igusa is on the Civil Engineering faculty at Northwestern University, Evanston, Illinois, USA. Professor Igusa has been working on problems related to reliability of civil engineering systems under random loads. He has recently been working on the relationships between chloride permeability of concrete and the statistical characterization of microstructure.

Professor S P Shah is Director of the NSF Science and Technology Center for Advanced Cement-Based Materials (ACBM), Northwestern University. Professor Shah's interests span a wide spectrum of current topics related to cementitious materials. During his first five years as Director of ACBM, he supported a series of innovative, basic research projects designed to answer fundamental questions with a strong emphasis on developing a materials science approach to studying cementitious materials. While this basic research is continuing, Professor Shah has also initiated a major shift in focus at ACBM toward the study of cement-based materials as a system and their use in society. Of primary interests are the related issues of improving environmental performance and long-term serviceability.

INTRODUCTION

Concrete technology has advanced dramatically in many fronts during the past few decades. Water-reducing admixtures make it possible to easily produce concrete with strengths over 100 MPa. Sophisticated structural analysis methods have facilitated the analysis of concrete structures with complex configurations and loads. Modern construction processes have reduced costs and increased quality of concrete structures. Thus, concrete has been, and continues to be, the material of choice for most of the world's transportation infrastructure.

While progress is continuing in all phases of concrete technology, the steady and highly visible deterioration of infrastructure has brought widespread scientific and engineering focus on long-term durability and service life prediction. The main issues in concrete durability are related to chemical and physical attack, either on the exposed surface or within the concrete material. Chemical attack has a variety of forms, including chloride ion permeation and the resulting corrosion of reinforcing steel, sulfate attack, carbonation, alkali-aggregate reaction, and leaching and efflorescence. Physical attack includes freeze-thaw cycles, wetting and drying cycles, wear and abrasion and temperature fluctuations [1]. The resulting deterioration, unfortunately, also has a variety of forms, including cracking, spalling, and loss of reinforcement bond.

These forms of attack and their influence on durability are highly interrelated, making research on concrete durability a challenging task. Herein, a discussion of some of the important issues on service life prediction with emphasis on concrete durability is presented. To limit the scope, the discussion does not go beyond the three major forms of degradation: chloride permeability, freeze-thaw cycles, and reinforcing steel corrosion.

RESEARCH OBJECTIVES

Research Focus

A major research goal is to predict the long-term performance of a concrete system using data collected during the material design and placement phases, with focus on the interactions between chloride permeability, freeze-thaw, and corrosion. The scientific challenge is significant since expected lifetimes of 50, 100, or more years far exceeds the short-term time available for materials testing. Furthermore, it is necessary to go beyond established long-term reliability techniques which are based on universally applicable extreme-value theories. Herein, two basic research areas are addressed: the development of destructive and non-destructive accelerated testing methods and the development of models based on materials and statistical sciences.

Materials-Based Durability Models

There has been a major research effort at the NSF Center for Advanced Cement-Based Materials (ACBM) and at other research institutions in the development of analytical and computational models for cementitious materials. These models describe the multiple length scales in concrete micro- and meso-structure, and include processes such as the transport mechanisms for chloride ions and other aggressive chemicals.

For cementitious materials applications, the system loads contain wide variability and extremes – thus future performance must be expressed in statistical terms. There are statistical techniques in "reliability engineering" to extrapolate long-term performance predictions from short-term data [2]. However, most of these techniques are generic and work only for systems with available long-term performance data. Thus, the theory would not be applicable to new testing methods. The view herein is that a synergy of materials and statistical sciences is the best approach for analyzing the durability of complex materials such as concrete. Materials science would be used to identify key observables that can be used as indicators of durability. These observables can be of a variety of forms, including direct measurements such as temperature and chemical concentrations, and cause-and-effect measures such as the behavior under short-term loads. Emphasis on durability will be on deterioration and other degradation mechanisms from the micron to the centimeter length scales. The materials-based reliability science would guide and be guided by innovative experimentation described in the next subsection.

Science-Based Short-Term Tests

We need to go beyond established reliability techniques based on accelerated testing [2] to extrapolate durability from short-term data. The usual tests involve relatively high-speed repetition of exaggerated magnitude loads such as temperature fluctuations, stresses, or chemical gradients and concentrations. The results of such tests would be used to predict the behavior of the structure under similar loads at much longer time scales and lower magnitudes. In the materials-based durability science, the various components of loads such as stress, surface attack, freeze-thaw cycles, and chemical attack would be integrated in deterioration models with multi-scale dimensions. With this scientific view of the cementitious material, an innovative battery of short-term tests would be developed based not only on duplicating long-term loads at accelerated pace and magnitude levels, but also on the integrated relationships between observables and performance. Destructive tests would be used for test specimens extracted during concrete placement. Non-destructive tests would be needed to experimentally investigate the relations between observables and deterioration for *in situ* specimens. Both types of tests are also needed to verify, calibrate, and guide the development of the materials-based durability models.

The observables of interest would be very broadly defined in materials science terms. It may be that an important observable may be one that has not yet been experimentally measured with precision. A current set of examples of such observables is associated with the micron-scale interface transition zone surrounding aggregate surfaces. The challenge here is to explore innovative experimental methods at various length scales to measure the critical observables. While the needs here are driven by the materials science – it may also be that advanced sensors developed for other applications may yield important observables in cementitious materials. In either case, the experimental data obtained from such sensors would be used in statistical methods to bring new insights in the materials-based durability science.

Impact on Materials Design

The testing methods and the approach to reliability analysis would be applicable to new designs of cementitious materials. A successful design of cementitious composites with targeted properties requires understanding and control of its inherently variable structure over a wide range of length scales.

Furthermore, to complete the materials design in a reasonable time frame, it is necessary to develop confidence in future long-term performance based on observations at short-term and medium-term time scales. Thus, research on service life prediction for durability would be vital in guiding the materials design process.

Impact on Analysis of Life-Cycle Costs

Predictions of service life based on durability analysis would be of direct and immediate use in any analysis of life-cycle costs. Due to the immense size of transportation infrastructure, a careful life cycle cost analysis of concrete systems would be a valuable tool in reducing monetary costs, environmental wastes and loads, and transportation congestion resulting from repair and replacement. A life-cycle cost of a concrete structure or system would guide the choice of materials, number of sensors, maintenance schedule, and replacement versus repair decision. In each phase of life cycle analysis, the state of the concrete material, as measured from sensors and non-destructive tests, and the associated prediction of long-term performance, would be needed. Longevity can be enhanced by well-timed, maintenance based on the observed material state. Even minor or localized damage should be detected and characterized in a timely and reliable fashion and repairs should be implemented in an optimal time frame, well before the costs of repairs would require unnecessarily high levels of expenditure.

POSSIBLE RESEARCH APPROACHES

Experimental Program – Direct Methods

The identification of the key observable parameters of concrete which affect long-term durability is an important research goal. Due to the complexity of cement and concrete, and the relatively short history of the materials science of cementitious materials, the identification process must rely heavily on experimentation. Since it is difficult to prepare and test many specimens, and since there are many parameters in concrete composition, statistical techniques are needed to maximize the information that can be gained from a limited data. Concepts from statistical experimental design [3, 4, 5] will be modified and used in the planning stages of the experimental program and in the analysis of the results. For each series of experiments, the measured responses of the specimens would be affected by the experimental parameters as well as by random variations associated with sampling and measurement. The former effect is commonly analyzed by response surface methods [6] and is based on estimates of the main and interaction effects of the parameters on the measured responses. The random effects are the identifiable components of the random variation of the responses, and are commonly called variance components [7].

Estimates for the variance components can be obtained by the method of moments (ANOVA) or by maximizing likelihoods. Using chloride permeability tests as a case study, Ankenman, et al. [8] have developed an approach to experimental design which allows for efficient estimation of both the main and interaction effects of the factors as well as the random variance component effects. This appears to be a promising approach for future research in more complex, interrelated degradation mechanisms.

The novel statistical design approach described above permits inclusion of more than one degradation mechanism in an experimental program. By performing a separate experimental series for chloride permeability, freeze-thaw, and reinforcement corrosion, the important relationships between these mechanisms could not be easily analyzed. It is the interactions between these mechanisms that are critical in long-term performance of concrete. For example, since the rate of corrosion is dependent on the ease with which ferrous ions and oxygen can diffuse or flow through concrete, considerable interaction with chloride permeability is expected. The aforementioned experimental design approach provides a systematic, statistically consistent strategy for studying such interactions.

The experimental strategy for analyzing interactions immediately dictates an important specification for the physical tests. The tests for chloride permeability, freeze-thaw, and corrosion must be performed on the same specimen. This precludes certain standard accelerated tests such as the AASHTO T277 rapid chloride permeability test [9, 10] and the ASTM C666 freeze-thaw test [11] since they cause significant, irrevocable changes to the specimen. Thus, it will be necessary to use or develop non-destructive and possibly relatively slow testing methods for the three degradation mechanisms.

Since the specimens must include steel reinforcement to examine corrosion, electrochemical methods for accelerated testing of chloride permeability [12] would be inappropriate. Hence, it will be necessary to use substantially slower diffusion tests, which are variations of the standard ponding test. Freeze-thaw cycling may still be applicable, and there are two testing options to be considered. Freeze-thaw cycling can be performed simultaneously with chloride diffusion, provided that the diffusion-testing device can be placed in the freeze-thaw apparatus. In this case, accelerated thawing with water baths could not be used – however, the length of time required for the diffusion tests makes this accelerated procedure unnecessary. If freeze-thaw is not performed simultaneously with chloride diffusion, then the specimen could be subjected to the accelerated procedure before the lengthy diffusion test. As for the corrosion test, the chloride ion solutions needed for the chloride diffusion tests would automatically accelerate the corrosion process by chemically destroying the passive oxide layer around the reinforcing steel [1, 13]. Specifically, the threshold at which this initiation of the corrosion process occurs is when the chloride concentration in relation to the hydroxyl ion concentration (Cl^-/OH^-) is approximately 0.6 [14].

While the standard accelerated tests for chloride permeability and freeze-thaw could not be used directly to study interactive effects, it is acknowledged that they are vital in many practical applications. Thus, there is also a need to determine the relationships between measurements from accelerated and non-accelerated tests. This, again, is also determined using statistical techniques of calibration.

The concepts described for studying interaction effects would be useful in the analysis of other multiple-degradation mechanisms of concrete. The development of specialized sensors for obtaining data for such multiple mechanisms as well as physical models for understanding interaction effects appears to be progressing rapidly [15, 16, 17, 18, 19, 20, 21].

Experimental Program – Indirect Methods

Indirect methods for testing and analyzing concrete are becoming accepted, although the underlying mechanisms between cause and effect still need further investigation. To be effective, indirect experimental tests must be combined with physical models for the degradation mechanisms.

The chloride permeability of concrete is generally evaluated by direct measurement of the diffusion coefficient, by monitoring chloride profiles moving into a concrete section, or by indirect methods such as the electrical conductivity test AASHTO T277, commonly known as the Rapid Chloride Permeability Test. Since, direct diffusivity methods take a long time for completion, indirect accelerated methods such as AASHTO T277 which measure ionic conduction under the application of electric field are used. The relationships between the direct and indirect tests would have to be experimentally calibrated, and further investigated by analytical and numerical models.

As for physical damage caused by freeze-thaw loads or reinforcement corrosion, indirect non-destructive testing methods will be needed. Damage may take the form of deteriorated mechanical properties, discrete defects such as cracks, regions of damaged material, or reduced bonding between any interfaces. The use of ultrasonic and sonic stress wave-based methods enables deep penetration and offers direct information concerning the elastic moduli, or other constitutive parameters, and presence of flaws. However, the current state of the art is still in the developmental phase. Advances have to be made to generate suitably controlled signals in concrete, to effectively detect signals, and to interpret quantitative measurements for evaluation of various aspects of degradation [22, 23, 24, 25, 26].

An important part of the research work is the development of new equipment and suitable techniques for specific application to concrete system testing. An effective use of ultrasound/sound in this regard will require the following: a device to generate repeatable and controllable ultrasound/sound signals in concrete structures; a technique for far-side (bottom) crack detection in pavements; and a technique for characterization of interfacial properties. The stress wave source will most likely be mechanical, as opposed to piezoelectric, in nature. A pneumatic source or an electric impact gun may provide repeatable and controllable signals. Emphasis will need to be placed on tailoring the sources to the various developed non-destructive stress wave techniques. Thus, the signal source will have to be variable in terms of frequency content and generated directivity field. The source must also be applicable to relatively rough surfaces, such as those found on unprepared airport and highway pavements.

Modeling Program – Direct and Indirect Methods

Ionic conduction through concrete depends on its microstructure (10^{-6} m) and mesostructure (10^{-3} m). Conductive microstructure includes gel and capillary pores in cement paste, voids in the weak matrix-aggregate interfacial zone and microcracks due to volumetric changes. Mesostructure, which includes aggregates, air voids, and macrocracks, also influences conductivity by providing obstacles to the macro flow of ions. Hence, conductivity in concrete is characterized by wide range of feature sizes, from nanometer pores to centimeter size aggregates. For any study of transport properties of concrete to be valid, conductivity of samples should be large compared to the length scale (correlation length) of the heterogeneities so that average flow rates can be used to analyze the system [27].

The conductivity of porous media, such as concrete, has traditionally been described using continuum and discrete models. Continuum modeling is the classical engineering approach for describing complex materials like concrete, characterized by several length scales. Effective properties of a porous medium are defined as the averages of corresponding microscopic quantities.

The averages must be taken over a volume small enough compared to the volume of the system, but large enough for applicable equations of change (e.g., Darcy's law and Fick's Law) to hold when applied to that volume. Predictions are difficult due to the complex nature of the system (length scales and interconnectivity) and many empirical factors are often added to account for disparities between theoretical and experimental results. Discrete models on the other hand do not suffer from these limitations because the description can be anywhere from sub-microscopic level to megascopic levels. These models are based on a discretized description of the material in terms of networks. The only difficulty is the amount of computational effort required and knowledge of applicable mathematical equations. Some of the ideas applied included percolation theory, diffusion processes, fractal concepts and universal scaling laws, which describe how and under what conditions the effective macroscopic properties of the system may be independent of its microscopic details [27].

The effective medium theory models are continuum models based on the assumption that effective transport properties of a disordered medium can be replaced by a hypothetical homogeneous one which mimics the behavior of the disordered one and then calculate the homogeneous medium's properties. In case such a substitution is possible, the problem of calculating the effective transport properties of the homogeneous medium is much simpler than the original problem as this is only an approximation of the original problem. Historically, there have been two approaches to this problem. The *t-matrix* and *non-self-consistent approximation* is applicable in the case when isolated inclusions are embedded in a continuous matrix consisting of a single phase and the properties of the effective medium are obtained by placing a sphere (or an ellipse) of the homogeneous or the effective medium in this matrix [28]. In the second approach [29], each homogeneity is embedded in the effective medium itself, the unknown properties of which are determined in such a way that the volume average over all the inhomogeneities yields no extra fields in the medium. This is usually called the effective-medium theory (EMT) and is the focus of this paper. Garboczi, et al., [30] provide a comprehensive account on the application of the theory to cementitious materials.

There exists a vast literature on predicting the effective conductivity of the two-phase composite materials [31, 32]. Virtually all of these models use perfect interface between the constituents, which is not the case in reality and especially in concrete. Recently, Torquato & Rintoul [33] used classical minimum energy principles to find very sharp, rigorous bounds on the effective properties of a class of composites with imperfect interfaces that incorporate crucial microstructural information at the interface. Cheng & Torquato [34] have proposed an extension of the Rayleigh's dipole model to a two-phase media with a super thin interface of different properties than other two phases. It is difficult to apply such models to concrete because of the randomness of the material and difficulty of computation. The EMT models are very suitable to the conduction problem in concrete because of its basis in electric fields and averaging approximations involved.

The history of the application of the EMT in concrete is very short. Garboczi and Bentz [35] were the first to apply EMT to mortars in calculating the effective conductivity of a three phase composite consisting of aggregate, interface transition zone (ITZ) and matrix. They developed formulations for self-consistent (SC-EMT) and differential (D-EMT) methods. The SC-EMT has roots going back to Bruggeman [29]. In the D-EMT the exact solution for very low volume concentration can be used to generate an approximate equation that can be solved for the effective conductivity.

The equations are very complex and require the use of approximations for several gradings [30]. The general effective medium theory proposed by McLachlan [36] has been applied by Christensen, et al., [37] to conductivity of paste.

However, relatively little work has been done on how the conductivity of concrete depends on quantities such as volume fraction and size distribution of aggregates and on the geometry and properties of the cement paste-aggregate interface and dependence of such properties on the statistical nature of concrete.

It is believed that a formulation based on EMT and statistical science will provide useful insight to the problem. The input to the model should be mix design variables such as volume fraction of aggregate, gradings, and water/cement ratio. Due to the highly subjective nature of the determination of thickness and conductivity of the interface, these quantities are best determined from experimental data using non-standard statistical techniques.

The techniques are non-standard because the interface properties cannot be measured or deduced directly. An inverse problem, based on inverting and statistically analyzing an analytical model with respect to experimental data, must be solved. To illustrate the analysis procedure for combining an analytical model with statistical techniques, an extended example will be briefly reviewed in the next subsection.

Modeling Program – Example of an Inverse Analysis

One important measure of concrete durability is the diffusivity of chloride ions. A standard measure of this diffusivity is the electrical conductivity test AASHTO T277, in which ionic transport is accelerated by an applied electric field. The ionic transport is primarily through an interconnected pore network and is controlled by the amount of evaporable water present in the paste, and concentration of the mobile ions in the pore fluid [12]. While the aggregate phase is relatively impermeable, it has a significant influence on conductivity because of the effects of dilution, tortuosity, interfacial transition zones (ITZ), and percolation [14].

Dilution refers to the decrease in the conductivity of a composite due to the addition of aggregates, which act as impermeable inclusions [30]. A related effect is the tortuosity of the path of the ions around the aggregates that leads to decreased ionic transport. The formation of the ITZ is due to the so-called wall effect, which results in an area of variable width around the aggregates with greater porosity, conductivity and w/c ratio than the surrounding matrix [38]. Finally, percolation refers to the connectivity of ITZ's at critical volume fractions that results in highly conductive preferential pathways [39].

This example illustrates the following modeling goals:

1. Systematic identification of the significant variables affecting chloride transport;
2. Study of the effects of ITZ conductivity and porosity;
3. Modeling of the effects using statistical and material science principles.

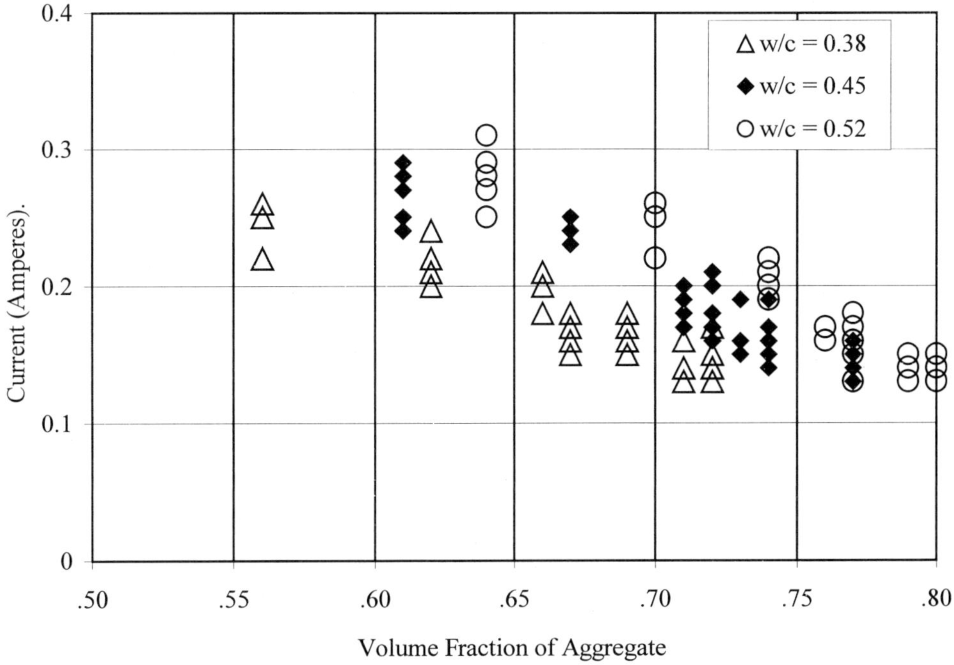

Figure 1 Relation between measured current and volume fraction of concrete specimens

The experimental design was based on a fractional factorial design with w/c = 0.38, 0.45, and 0.52, four aggregate gradings, and two maximum aggregate sizes [8]. The effect of aggregate volume fraction and w/c on conductivity (as measured by current) is shown in Figure 1. The results indicate that increasing the volume fraction of aggregate and decreasing the water-to-cement ratio decreases the concrete conductivity. These relations are already well known. While a direct statistical analysis, using standard regression, can be applied to fit curves to the data, the results would not be tied with a materials science model. Furthermore, there would be no systematic means for qualitatively or quantitatively assessing the effects of ITZ conductivity and porosity. To analyze the experimental results further, models based on materials science concepts are needed [40].

One widely used equation is of the form:

$$\sigma = \sigma_p (1 + m [VF])$$

where: σ and σ_p are the conductivity of the composite (concrete) and matrix (paste), respectively, the coefficient m indicates the first-order effect of the aggregate on the composite conductivity, and [VF] is the volume fraction of aggregate. This equation is valid in the dilute limit, where [VF] is of the order of 20%, although it has been used with some success at higher [VF]. Maxwell was the first to use this equation, and for spherical aggregate he obtained $m = -1.5$.

The negative sign is due to the dilution effect, where increased aggregate volume fraction decreases the paste volume fraction resulting in reduced conductivity. The magnitude of *m* is larger than unity because of the tortuosity effect, where the tortuous paths of the ion flow leads to further reduction in conductivity.

Garboczi, et al., [30] generalized the dilute limit equation in two ways: they considered multiple aggregate sizes (grades) and the effect of a conductive ITZ. Their result is:

$$\sigma = \sigma_p \sum_i (1 + m_i [VF_i])$$

where: $[VF_i]$ is the volume fraction of aggregate grade *i* and:

$$m_i = 3\alpha_i \frac{2(\alpha_i - 1)\sigma_{ITZ} - (2\alpha_i + 1)\sigma_p}{2(\alpha_i - 1)\sigma_{ITZ} + 2(2\alpha_i + 1)\sigma_p} \qquad \text{where } \alpha_i = \left[1 + \frac{\text{ITZ thickness}}{\text{average radius of grade } i} \right]^3$$

in which: σ_{ITZ} is the conductivity of the ITZ. A thorough interpretation of m_i is given in Garboczi, et al., [30]; in short, m_i becomes less negative (and may become positive) as the relative conductivity and thickness of the ITZ increases. An alternate formulation of the composite conductivity is based on Bruggeman's EMT results [29], with percolation effects by Mclachlan, et al., [36]:

$$\sigma = \sigma_p \prod_i \left(1 - \frac{[VF_i]}{[VF_c]} \right)^{-m_i [VF_c]}$$

where: $[VF_C]$ is the critical volume fraction corresponding to percolation. For this example analysis, a simple model for the ITZ porosity is used, where the ratio of ITZ and paste porosities is specified by a constant, β.

An inverse relation is obtained using the aforementioned model with the experimental data as follows. In Figure 1, the independent variable was the aggregate volume fraction, [VF], which is directly related to the experimental design parameters; the dependent variable is the conductivity of the concrete, σ, which is directly related to the data measurements. The problem with this choice of variables is that there is no one-to-one functional relation between conductivity and aggregate volume fraction, as clearly indicated in the figure. In the inverse analysis, the conductivity of the paste, σ_p, is chosen for the dependent variable and the w/c ratio for the paste is chosen for the independent variable. A one-to-one functional relation between these variables is expected because the relation is an intrinsic property of the paste, independent of the aggregate. However, since neither the conductivity of the paste nor the w/c ratio for the paste can be directly or indirectly measured, they must be estimated by inverting the preceding analytical models. In the models, the experimental data is substituted for the conductivity of the concrete, σ, and the parameters for the ITZ, including its thickness, relative conductivity, σ_{ITZ}/σ, and relative porosity, β, are obtained by statistical means. Two results are shown in Figure 2. In Figure 2(a), the relation between paste w/c ratio and conductivity is plotted without including ITZ in the analytical model, and in Figure 2(b), the same relation is plotted with ITZ thickness of 0.02 mm, relative conductivity $\sigma_{ITZ}/\sigma = 6$, and relative porosity, $\beta = 1.7$.

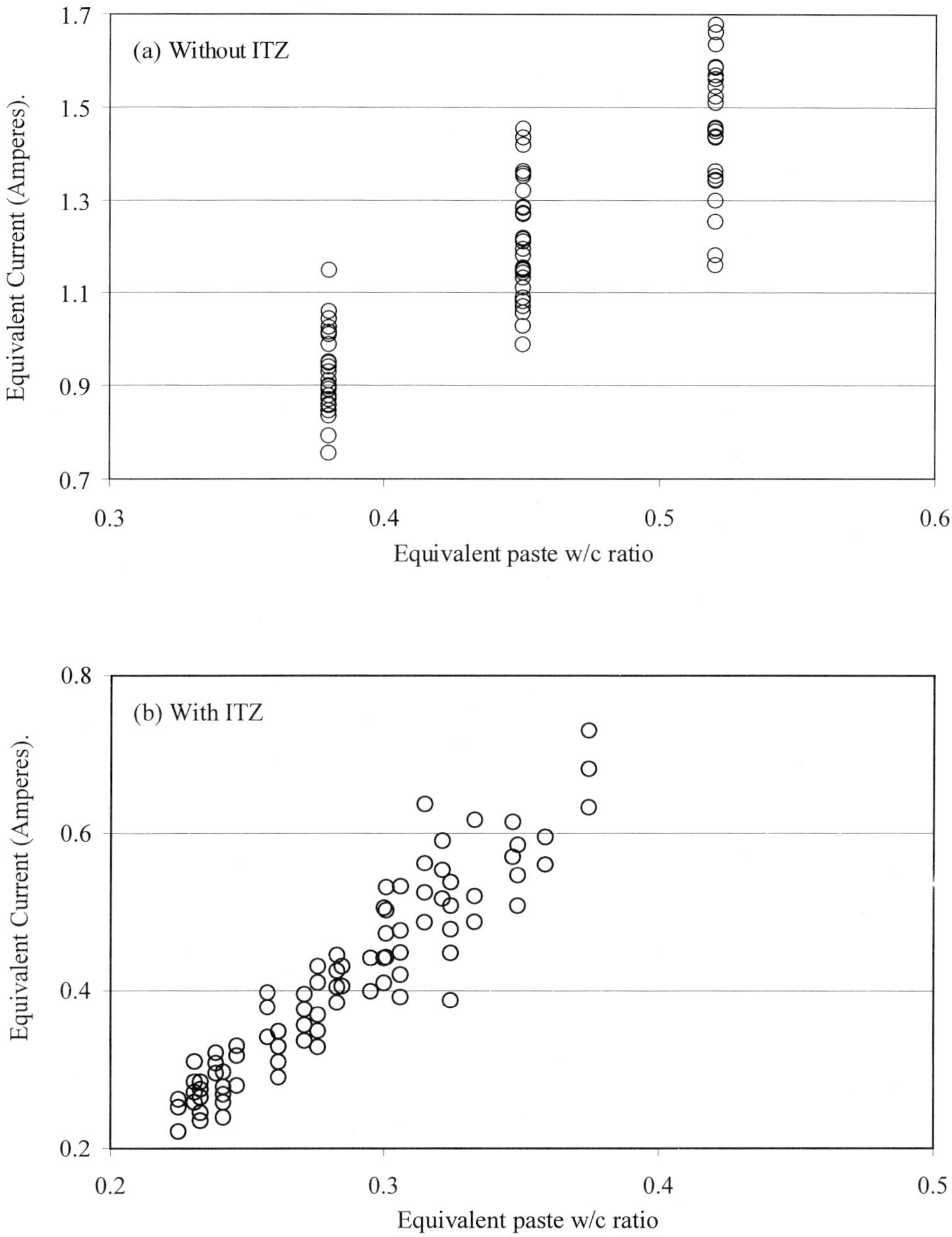

Figure 2 (a) and (b) showing inferred relations between cement paste w/c ratio and conductivity

It can be seen that the ITZ model provides a clearer functional relation between the two variables, while the model without ITZ is widely scattered. Improvements in the modeling are needed as the paste w/c ratios are quite low. Nevertheless, the results for the ITZ parameters are consistent with other results reported in the literature [30]. Furthermore, this example analysis provides indirect evidence of the importance of the ITZ in concrete conductivity.

CONCLUDING COMMENTS

In future work on service life prediction with emphasis on concrete durability, two main research areas appear to be important. One is in developing integrated testing methods for testing the ability of a concrete specimen to withstand chloride penetration, freeze-thaw cycles, reinforcing bar corrosion, and other degradation mechanisms. The other is in developing models of varying complexity that can be used to interpret the experimental results and produce quantitative assessments of concrete durability. A balance must be maintained between these two research areas, since models with sufficient experimental justification and testing procedures with sufficient materials science backing would provide the desirable fundamental advances in the state-of-the-art.

ACKNOWLEDGEMENTS

The authors would like to express their appreciation for the research support by the program DMS-9313013 from the U.S. National Science Foundation.

REFERENCES

1. MINDESS, S AND YOUNG, J F. Concrete, Prentice Hall, 1981.

2. NELSON, W. Accelerated Testing: Statistical Models, Test Plans, and Data Analyses. Wiley, New York, 1990.

3. COX, D E. Planning of Experiments. Wiley, New York, 1958.

4. BOX, G E P, HUNTER, W G AND HUNTER, J S. Statistics for Experimenters. Wiley, New York, 1978.

5. MEAD, R. The Design of Experiments. Cambridge University Press, Cambridge, 1988.

6. MONTGOMERY, D C AND MYERS, R H. Response Surface Methodology. Wiley, New York, 3rd edition, 1995.

7. SEARLE, S R, CASELLA, G AND MCCULLOCH, C E. Variance Components. Wiley, New York, 1992.

8. ANKENMAN, B, KARR, A F, LIU, H AND PICKA, J D. Experimental Design for Estimating a Response Surface and Variance Components. Technometrics, submitted paper.

9. AASHTO. Standard method of test for rapid determination of chloride permeability of concrete. American Association of State Highway Transport Officers, T277-83, 1983.

10. WHITING, D. Rapid determination of the chloride permeability of concrete. Final Report FHWA/RD, NTIS PB82130724, Washington DC, August, 1981.

11. ASTM. Standard Test Method for Resistance of Concrete to Rapid Freezing and Thawing, ASTM Standard C666, Annual Book of ASTM Standards, Vol. 04.02, pp. 322-327.

12. GJØRV, O E AND ZHANG, T. An Electrochemical Method for Accelerated Testing of Chloride Diffusivity in Concrete. Cement and Concrete Research, Vol. 24, No. 8, 1994, pp. 1534-1548.

13. RILEM. Corrosion of Reinforcement and Prestressing Tendons: A State of the Art Report. Materials and Structures/Research and Testing, Vol. 9, No. 51, 1976, pp. 187-206.

14. HILSDORF, H K. Concrete Compressive Strength, Transport Characteristics and Durability. Performance Criteria for Concrete Durability, RILEM Report 12, Eds. Kropp, J and Hilsdorf, H K, pp. 165-195.

15. DESOUZA, S J, HOOTON, R D AND BICKLEY, J A. A Field Test for Evaluating High Performance Concrete Covercrete Quality. Canadian Journal of Civil Engineering, Vol 25, No. 3, 1998, pp. 551-556.

16. CABRERA, J G, CLAISSE, P A AND LYNSDALE, C J. Application of Microcomputers in the Retrieval and Processing of Data in the Study of the Durability of Reinforced-Concrete. Materials and Structures, Vol. 26, No. 164., 1993, pp. 587-593.

17. JONES, R, CHIU, W K, PITT, S AND GALEA, S C. Structural Integrity Monitoring. Engineering Failure Analysis, Vol. 4, No. 2, 1997, pp. 117-131.

18. AMEY, S L, JOHNSON, D A, MILTENBERGER, M A AND FARZAM, H. Predicting the Service Life of Concrete Marine Structures: An Environmenal Methodology, ACI Structural Journal, Vol. 95, No. 2, 1998, pp. 205-214.

19. MARCHAND, J, SELLEVOLD, E J AND PIGEON, M. The Deicer Salt Scaling Deterioration of Concrete – An Overview, Ed. Malhotra, V M, ACI Special Publication SP-145, 1994, pp. 1-46.

20. BILODEAU A AND CARRETTE, G G. Resistance of Condensed Silica Fume Concrete to the Combined Action of Freezing and Thawing Cycling and Deicing Salts, Ed. Malhotra, V M, ACI Special Publication SP-114, 1989, pp. 945-969.

21. MONTEIRO, P J M, GJØRV, O E AND MEHTA, P K. Microstructure of the Steel-Cement Paste Interface in the Presence of Chloride, Cement and Concrete Research, Vol. 15, 1985, pp. 781-784.

22. MAJI A AND SHAH S P. Process Zone and Acoustic Emission Measurement in Concrete, Experimental Mechanics, 1988, pp. 27-33.

23. LANDIS, E N, SHAH, S P AND LI, Z. Application of Quantitative NDE to Basic Fracture Research of Concrete, Review of Progress in QNDE, Vol. 13, 1994, pp. 2123-2130.

24. OHTSU, M, OKAMONOT, T AND YUYAMA, S. Moment Tensor Analysis of Acoustic Emission for Cracking Mechanisms in Concrete, ACI Structural Journal, 1998, pp. 87-95.

25. HEARN, S W AND SHIELD C K. Acoustic Emission Monitoring as a Nondestructive Testing Technique in Reinforced Concrete, ACI Structural Journal, 1997, pp. 510-519.

26. SHAH, S P, SWARTZ, S E AND OUYANG C. Fracture Mechanics of Concrete: Application of Fracture Mechanics to Concrete, Rock and Other Quasi-Brittle Materials, John Wiley & Sons, Inc., New York, 1995.

27. SAHIMI, M. Flow and Transport in Porous Media and Fractured Rock. VCH, 1995, pp. 1-10 and 51-53.

28. MAXWELL, J C. Electricity and Magnetism, 3rd edition, Oxford, 1892, pg. 440.

29. BRUGGEMAN, D A, Annals of Physics, Vol. 24, 1935, pg. 636.

30. GARBOCZI, E J, SCHWARTZ, L M AND BENTZ, D P. Modeling the Influence of the Interfacial Zone on the DC Electrical Conductivity of Mortar. ACBM Journal, Elsevier Science, 1995.

31. RAYLEIGH, LORD, On the Influence of Obstacles Arranged in Rectangular Order upon the Properties of a Medium. Philosophical Magazine, Vol. 34, 1892, pp. 481-502.

32. HANSHIN, Z. Analysis of Composite Materials, Journal of Applied Mechanics, Vol. 50, 1983, pp. 481-505.

33. TORQUATO, S AND RINTOUL, M D. Effect of the Interface on the Properties of Composite Media. Physics Review Letters, Vol. 75, 1995, pp. 4067-4070.

34. CHENG, H AND TORQUATO, S. Effective Conductivity of Periodic Arrays of Spheres with Interfacial Resistance. Proceedings of the Royal Society of London A, Vol. 453, 1997, pp. 145-161.

35. GARBOCZI, E J AND BENTZ, D P. Analytical Formulas for Interfacial Transition Zone Properties. Materials Research Society, Fall 1996, preprint.

36. MCLACHLAN, D S, BLASZKIEWICZ, M AND NEWNHAM, R E. Electrical Resistivity of Composites. Journal of American Ceramic Society, Vol. 73, No. 8, 1990, pp. 2187-2203.

37. CHRISTENSEN, B J, COVERDALE, R T, OLSON, R A, FORD, S J, GARBOCZI, E J, JENNINGS, H M AND MASON, T O. Impedance Spectroscopy of Hydrating Cement-Based Materials: Measurement, Interpretation, and Application. Journal of American Ceramic Society, Vol. 77, No. 11, 1994, pp. 2789-2804.

38. OLLIVIER, J P AND MASSAT, M. The Effect of the Transition Zone on Transfer Properties of Concrete. Interface Transition Zone in Concrete, Ed. J C Maso, RILEM Report 11.

39. WINSLOW, D F AND COHEN M D. Percolation and Pore Structure in Mortars and Concrete. Cement and Concrete Research, Vol. 24, 1994, pp. 25-37.

40. JAISWAL, S, IGUSA, T, STYER, P, KARR, A F, AND SHAH, S P. Influence of Microstructure and Fracture on the Transport Properties in Cement-Based Materials. Brittle Matrix Composites, International Conference on Brittle Matrix Composites, Eds. Brandt, A. M., et al., Warsaw, Poland, October 6-8, 1997, pp. 199-220.

INTEGRATING DURABILITY INTO THE DESIGN PROCESS

G C Hoff

Mobil Technology Company

United States of America

ABSTRACT. This paper discusses an approach that should be followed if the integration of durability design into the structural design process is to be successful. The inclusion of durability requirements in national codes and standards is probably the surest way to get uniform acceptance of these requirements by both designers and constructors. There is a definite need for shorter code and standard revision cycles, however. More definition of the environment in which the concrete structure will perform is necessary. Realistic service life requirements are necessary for every structure. Performance criteria should focus on the transport properties of the concrete, as it is the ingress of the environment into the concrete that is the major problem. Combinations of both performance and prescriptive specifications are necessary to insure adequate durability. The evaluation of concrete performance should use national standards whenever possible rather than the vast array of tests proposed by many different researchers. Changes in the education process, both at the university level and for experienced engineers, would help remove resistance to changes in the design process. The need to include the durability considerations into the project planning is essential. Closer collaboration of the designers and the constructors with a view to eliminating structural details that can lead to durability problems is needed. The need for rigorous quality control and quality assurance on projects is essential to insure satisfactory durability in the future.

Keywords: Codes, Concrete, Constructability, Detailing problems, Durability criteria, Education, Environment, Project timing, Service life, Specifications, Standards, Structural design, Test methods.

Dr George C Hoff is the Senior Concrete Technologist for the Mobil Technology Company, Dallas, Texas, USA. He specializes in concrete for marine and offshore concrete structures and has done extensive work with lightweight aggregate concrete, and high strength/high performance concrete. He is a past-President of the American Concrete Institute (ACI) and the current Chairman of federation internationale du beton (*fib*) Commission 10 on Construction.

INTRODUCTION

Concrete is the most extensively used material in the $400 billion per year construction industry in the U.S. It is also the most extensively used material in most other regions of the world. This common use contributes to the continuing problem of poor concrete durability as nobody wants to pay the extra attention and the little extra initial cost it takes to provide more durable, longer lasting concrete. Because we use so much concrete, and because concrete has competing materials (eg., steel, wood, asphalt) for many applications, the tendency is to provide the concrete at the lowest possible cost. This is achieved by minimizing the materials required to produce the concrete, using the cheapest (and probably the least trained) labor, and eliminating the good practices (eg., quality control, consolidation, curing) that are necessary to produce good concrete.

What does this type of approach bring us? A 1990 National Research Council report in the U.S. indicated that it will cost $2 to $3 trillion over the next 20 years to repair all the U.S. concrete structures which are now deteriorating from corrosion or are poorly made and maintained. This is not to say that the competing construction materials survive any better. In fact, most do worse than even the poorest of concretes. A long running television commercial in the U.S. promoting the sale of automobile oil filters, concluded by having the auto repair mechanic saying "You can pay me now or you can pay me later!". The inference was that if you initially paid a little more for the oil filter that you used in your automobile, you wouldn't have to pay for costly repairs in the future. A similar analogy exists for concrete.

The need for a more durable concrete has not been lost on the global engineering community. As a result, most countries now have their own programs to develop "high-performance concrete" because they recognize that there is a great potential to minimize expense each year in maintenance, repair and replacement of poorer quality concrete. What exactly is high-performance concrete?

Many definitions have been proposed for "high performance concrete." In general, high performance means behavior above and beyond what we would normally expect from concrete. This tends to be somewhat of an anomaly because when we order concrete from a supplier, we expect it to have satisfactory performance with respect to both strength and durability. We certainly don't order "low performance" concrete but that is often what we get. We typically order concrete by strength requirements. Strength is quantifiable. If we order 40 N/mm^2, we can measure it. How do we order durable concrete? If we ask for concrete that will last for 100 years, how do the concrete supplier or we know that it will? We won't be here to check and see that it has performed satisfactorily and, most likely, the concrete supplier won't be in business to compensate us for any non-performance if it fails in less than 100 years.

Indications are that high performance concrete is having a successful evolution, at least at the laboratory level. However, the money being spent on developing and evaluating high performance concretes is, for the most part, distributed among many researchers in many countries doing relatively small, uncoordinated studies. It is estimated that in the U.S. in 1998, there were 3000 concrete studies being done in 500 to 1000 laboratories or organizations. How this information gets disseminated and integrated into actual practice is a major problem.

Dissemination of information is a simpler problem than the integration of that information into the design process. Meetings of the type that this presentation is being made at, are helpful for information dissemination in that researchers are advising other researchers of their work and results. Normally researchers have trouble tracking what other researchers are doing as is evidenced by that fact that many papers on well researched durability problems (eg., reinforcing bar corrosion) reference only the work done on that subject by the author of that paper. However, the results from this meeting and the many similar meetings on concrete durability generally do not cause any significant changes in design or construction practice. Why is this? The following sections explore a few of the author's views on this problem.

It is also the author's belief that some of the major stumbling blocks to acceptance of durability considerations in design are education, timing of the introduction of durability considerations into a project, and the reluctance, by constructors, to change existing practices to accommodate durability requirements. All have a financial impact on a project as an insured durability comes at an initial higher cost. Some aspects of the influences of these topics are described below as catalysts for further discussion. Bryant Mather, world renowned expert on concrete durability, has often stated that based on what we know about concrete and its behavior in given environments, the only reasons concrete should deteriorate are ignorance, stupidity or fraud. At least 15 years ago, George Somerville [1], another world renowned expert on concrete durability, described what was necessary to provide long-lasting, durable concrete structures. Yet today, we are still struggling with the problem.

IMPLEMENTATION

Designers do not have a financial incentive to use the new technology being developed by the researchers on high performance concrete. Using or specifying new technologies tend to increase the designers cost and also the risk of doing business without any appreciable benefit or financial reward. A similar problem exists for the constructors as they generally are not a part of the design process and can only execute work in accordance with the documents of the contract. Occasionally, "value engineering" clauses exist in contracts which allow the constructor to use innovative construction materials and techniques to produce cost savings on a project. The owner and the constructor share these savings, but these contract provisions are rare and there are numerous critics of this approach. For a constructor to deviate significantly from standard practice to implement innovations increases his risk which is a strong disincentive when most contracts are awarded on a least price basis.

When the designer is reluctant to specify new technology and the constructor is reluctant to implement it as his own discretion because of risk or cost, the typical tendency is to think of the government, at all levels, as the solution to the problem. Governments are in a position to lead in the development and demonstration of improved concrete technology by implementing constructions that use technology that may be outside of the provisions of codes and standards. This is due to the fact that governments are self-insuring and that they are entrusted with providing safe structures for public use using the publics money.

However, in many governments, the motivation for providing long-lasting materials does not exist. The principal motivators in taking action are a fear of disaster (eg., a bridge collapse) and lowest cost. You cannot sell the politicians on the concept of durability because the material only has to last as long as their term in office. Life-cycle costs are generally not considered.

In private corporations, similar attitudes often exist because stockholders expect business to be performed at lowest costs. There are many examples of developers doing the minimum needed in construction because they do not intend to be the long term owner of the structure and hence do not care what happens to the structure in the years to come.

Would the designer, the constructor, and the government be more willing to adopt improved durability innovations if they could do it by reference to accepted codes, standards, and specifications? The answer is probably "Yes" because there is a general underlying faith by the industry and the general public in these types of documents. The codes and standards currently determine technology acceptance. If the accepted codes, standards, and specifications require it, then it puts all competing parties on a level playing field. Risks become the same and the durability innovations become part of the cost of doing business for everyone.

In support of this premise, the US Civil Engineering Research Foundation (CERF) conducted a survey of 600 members of the highway community, including public sector officials, construction companies, entrepreneurs, and research facilities to determine what were the most common barriers to adopting innovations in highway construction technologies. The ranking was 1) constrained standards or specifications (23%), 2) current limitations on proprietary product usage (18%), 3) lengthy processes or high costs (17%), and 4) known evaluation inadequate (12%). If the standards or specifications do not allow the use of new technology to improve the performance and service life of concrete, then that is a problem we certainly can fix.

THE STANDARDIZATION PROCESS

Structural design of concrete typically never considers the durability aspects of the concrete. Strength is the primary consideration along with modulus of elasticity and other related mechanical properties. The integration of structural design and durability design can be approached from two directions: higher strengths lead to improved durability or improved durability leads to higher strengths. Neither approach is universally correct and exceptions can always be found. From a practical standpoint, it may be simpler to concentrate on improving the durability aspects by changes in durability requirements of codes and standards. Code changes with respect to improved durability are typically accomplished in a more timely manner than changes to structural provisions. Many structural requirements in codes and standards were developed for concretes of strengths 40 N/mm^2 or less. When higher strengths are used, these requirements may or may not become more conservative. Changes in the structural provisions need supporting data from tests of structural members and the production of such data is a long and expensive process, more so than for durability evaluations. However, if improved durability design results in routinely occurring higher strengths, the designers will, in time, adjust their procedures to effectively use this additional strength.

It is the author's belief that standards must be the place to begin to integrate durability requirements into the design process as many specifications simply refer to the standards. As noted above, constrained standards or specifications are a significant barrier to the implementation of new materials and technology into the design and construction process. Why are they constrained or restrictive? The existing design codes and standards are usually conservative because their underlying basis is to protect the public while providing a functional structure. They tend to restrict the use of new materials until, and unless, a substantial database about them has been established. They are also flavored by the perception of what the industry can realistically produce. This is a dangerous perception because experience has shown that with a little training and education, high quality concrete production and construction can be achieved on a routine basis.

The preparation of codes, standards, and specifications are done differently in various parts of the world. Some countries use the "consensus standards" approach where representatives from all aspects of the industry, including academia, jointly prepare and approve the standards. Some countries use government agencies or teams of specialists to prepare the standards. Other countries simply adopt the standards of other countries and modify them for their local needs. All of these preparation methods run the risks of influence by special interest groups or individuals, or, even more commonly, having uninformed individuals either doing the standard preparation or approval or both. In general, these risks are reduced in the consensus standards approach because of the checks and balances provided by a very broad representation of the industry in the standards preparation.

How can we improve the standardization process to expedite durability requirements into the design process? The first step should be the timely and efficient evaluation of the durability data base being developed by the research and development community. This should be complimented by demonstration projects of the new technology when they exist. Unfortunately, when public funds are used, demonstration projects are not viewed favorably because they tend to benefit only a limited segment of the community instead of everyone who helped pay for it. The key word here is "timely". It is not very effective if the inclusion of new technology into standards takes 7 to 10 years, yet that time period is typical for many code and standard reissue cycles. A classic example is the preparation, in the US, of a standard for silica fume, a major contributing material for improving durability. After 11 years of development, Draft 17 of the standard was finally approved. Interim procedures for adopting new technology into standards must be invoked so that once a new technology or material has been satisfactorily demonstrated, it can appear in the standards within 12 to 18 months. This may be an impossible task in the short term but is something that needs serious consideration by the industry.

If we are to include more stringent durability requirements into our codes and standards, two approaches are possible: performance requirements and prescriptive requirements. Both have their advantages and disadvantages and perhaps combinations of the two can supply the best solution.

PERFORMANCE AND PRESCRIPTIVE SPECIFICATIONS

Performance specifications are promoted as providing flexibility and encouraging economic mixture proportioning. "Economic mixture proportioning" infers that either the least amount of material will be used in producing concrete that will satisfy certain performance criteria or that less expensive materials than the conventional concrete materials will be used. The establishment of "performance criteria" assumes that we understand the nature of concrete and its interaction with both the environment and service loads. It also assumes that we can adequately define the environment and also predict the service life of the structure. This is doubtful although we are making progress in all of these areas. Realistically, we are only making educated guesses for most aspects of "performance criteria". There is also an uncertainty about the methods we use to measure "performance". While the concrete technologist may be aware of these limitations, the concrete designer assumes that if the concrete meets the performance criteria set forth in standards or specifications, the concrete when used in a structure in a given environment will have sufficient durability for the service life of the structure. Unfortunately, it is very difficult to provide a general criteria which meets all possible uses of concrete.

Prescriptive specifications, those that tell you exactly how much of each ingredient to put into the concrete, used to be common and still exist in many parts of the world. The mixing water is typically the only variable and is adjusted to suit the conditions of the construction. This would be ideal if all the ingredients used to make concrete were the same worldwide. They are not, so the best we can do, based on our present research on concrete durability, is to define minimum amounts of binder material and maximum amounts of water to be used with that binder so as to provide a dense matrix to resist the intrusion of the environment. We can also require portions of that binder to be supplementary cementing materials, e.g., fly ash, silica fume, and blast furnace slag. These supplementary cementing materials further densify the binder. The use of chemical admixtures that allow the reduction of water while maintaining adequate workability can also be prescribed. These admixtures may be necessary when using some of the supplementary cementing materials. Prescriptive specifications tend to be in conflict with the concept of "economic mixture proportioning" for performance specifications but are more easily understood by the user.

It is doubtful that more durable concrete can be achieved by minimizing the amounts of the more expensive ingredients of concrete (namely Portland cement) while still obtaining adequate strengths. As noted earlier, concrete strength is, however, the primary consideration for the designer. Strength is, at best, an indirect measure of concrete quality from a durability viewpoint. Historically, most standards and specifications have used the water-binder ratio as the main parameter to define concrete for durability, with the 28-day compressive strength consistent with the water-binder ratio providing the field control of the concrete. For marine concretes, minimum amounts of cementing material are also specified. The extensive research being done on high-performance concrete, however, indicates that all these requirements for improved durability, while necessary, may not be enough. What is needed is a very dense matrix of ingredients that are compatible with the aggregates and resistant to the environment in which the concrete is to used. In this regard, the use of low water-binder ratios, high binder contents, and supplementary cementing materials (eg., silica fume, fly ash, blast furnace slag) in conjunction with Portland cement have typically shown superior performance to conventional concretes under many different environmental exposures. Most of these modifications to improve durability will not produce a more "economic" mixture. With the cost of the concrete materials being only 5 to 15% of the cost of the concrete in-place, it is also foolish to consider using insufficient amounts of material when considering what future maintenance costs will be.

THE ENVIRONMENT

There is one school of thought that believes that durability is not a property of the concrete but is a function of the environment. It postulates that all concretes are durable provided they are kept in an environment consistent with the composition of the concrete. Ancient concretes, made with natural cements, lasted for thousands of years in a benign environment. With the advance of civilization, the environment was changed as pollution increased and the concrete began to deteriorate. This can be witnessed in the ancient monuments in populated areas of Greece, Italy, Egypt and Mexico. Similar monuments in other regions of the world that have not seen significant pollution have not experienced this same level of deterioration. Concrete that survives in one environment may not survive in another environment. Concrete used in the interior of buildings may not survive when used in a marine environment, a sewage treatment plant, or for the containment of hazardous waste.

When concrete is put in a difficult environment, it will deteriorate at rates governed by that environment. If the rates of deterioration for that environment can be predicted, then sufficient material can be added to achieve a prescribed service life or the concrete constituents can be modified to reduce that rate and extend the service life. As an example, the corrosion of reinforcements has received extensive study for some time. Most of the principal factors that reduce the long-term integrity of reinforced concrete are well understood. The use of adequate concrete cover, dense concrete matrices, crack control measures, anti-corrosion admixtures and additives, and other techniques have resulted from this knowledge. Unfortunately, these are still relatively short-term solutions. Few studies have been made that can help predict the longevity of reinforced concrete over hundreds of years. Very little information in the literature is helpful in relating, in a quantitative way, the rate of degradation of concretes subjected to the principal characteristics of the environment over the long term.

Do we know enough about our environments to establish a universal criteria for performance specifications? Our specifications presently address marine environments, sulfate soils, freezing and thawing, and a few other phenomena which can be quantified. But what about pollution, acid rain, hazardous wastes and other environments. While the present answer to a universal criteria is probably "No", I am confident that, with time, individual criteria can be established for almost all aggressive environments. If we had this information today, we would still be at risk in designing structures with very long service lives because we don't know what our environment of the future will be. If we design concrete to last 1000 years based on our environment today, do we have any guarantees that the environment will not become worse in the future? Hopefully it won't, but we can not be sure.

SERVICE LIFE

Service life of engineering structures has been a hot topic within the engineering community for the last decade, but it still lacks adequate definition. In general, service life of a structure is a questionable concept. Very few concrete structures have a precise service life. Offshore concrete platforms have a service life that is determined by the amount of hydrocarbon in the reservoir it is producing and the rate of production. Yet, concrete platforms built in the 1950's for a service life of 20 years are still in service because the technology of extracting the hydrocarbons from the earth has improved over the years so that much more is being produced than was originally planned for. Buildings designed for 40 years are still in service after 70 years while similar buildings are removed in 10 to 15 years to make way for more modern buildings or other facilities.

For the sake of the designer, do we need to specify a service life for a structure? The answer is "Yes" even though we don't know how realistic it will be. Requiring that a structure should have a specific service life must also be accompanied by performance criteria that gives the designer a quantitative objective in the design process.

The service life of concrete is usually dependent on a slow rate of deterioration which results from combinations of the environment, the concrete quality, and the size and configuration of the concrete element. In addition, service life is also a function of the failure criteria which is used. A practical service life may also depend on the competition that concrete may receive from alternative materials. If a pavement is designed to last 100 years, it will probably never be built because the cost will not conform to most budgets of the agencies who fund highway construction. Cheaper materials than concrete will be used without regard to service life. Structures whose basic function and utility doesn't change with the passage of time, such as bridges, buildings, marine structures, dams and navigation structures, certainly should be designed to last at least 100 years. Other structures whose capacity will be exceeded within reasonable time periods because of increasing population or which will become obsolete due to changing technology, can have shorter lives. Whatever service life we specify, it should be on the conservative side because we can't anticipate all of the factors which might influence the structures longevity.

PERFORMANCE CRITERIA AND MEASUREMENTS

Although properties such as compressive strength, modulus of elasticity, and porosity are indirectly related to durability, the transport properties of diffusivity and permeability are probably the most important factors relating to durability. Unfortunately, these properties are not simple things to measure. The major degradation parameters that exist in a service environment include sulfate ions, chloride ions, leaching of calcium hydroxide by water, carbon dioxide reactions, alkali-aggregate reactions, freezing and thawing, wetting and drying, and these various parameters in combination. If the transport of these degrading phenomena into the concrete can be reduced or eliminated, the long-term performance of the concrete can be greatly enhanced. Only by establishing criteria which demonstrates that the environment is not penetrating the concrete at unacceptable rates, can we make an estimate of the service life of the concrete.

Having once established the criteria, we must have realistic methods to evaluate the concrete to ensure conformance to the criteria. Traditionally, concrete durability has been assessed by measuring either the strength, weight change, length change or change in structure by non-destructive techniques of specimens subjected to a corrosive agent or a specific difficult environment. The corrosive agents are generally applied at the external boundary of the specimen but may sometimes be included in the concrete during mixing. Difficult environments typically include extremes of temperature and varying moisture conditions which are varied throughout the test, usually for time periods which are more suited to the test procedure than the real world environment. Results from these types of tests may not be applicable to predictions of long term durability and are generally not sensitive to the design geometry of structural members.

The durability renaissance of the 90's has led to an abundance of performance test methods for specifying durability as concrete technologists have endeavored to move away from strength as a durability guide. In a quick scan of test methods related to the permeability or diffusion rates of concrete, the author found 11 different methods by which these phenomena could be quantified. Tests of this type that have been developed in the laboratory using a small range of trial concretes may not translate well to other laboratories, other concrete, or the variety of environments that exist in the real world. Test specifications that have not been formulated by national associations may be flawed in technique or description. In our specifications, we should use national standard tests wherever possible.

Initially, including durability performance criteria and the methods by which to measure them in the specifications will meet a great resistance from both the designer and the constructor who are not used to working towards a "durable structure". The criteria will be "too severe". The test methods will be "impractical". These are legitimate concerns and caution must be exercised to ensure that what is asked for is absolutely necessary to obtain a durable concrete structure. The inclusion of impractical criteria and test methods can be an unacceptable hindrance to construction that leads to excessive costs and delays. Apart from impracticality, some "durability tests" may be no better an indication of performance than strength. The sudden implementation of arduous performance specifications can be expected to cause significant start-up problems on jobs and may prove impractical to enforce. These are not insurmountable problems but will require some education of the industry during the implementation period.

EDUCATION

In the years to come, the integration of durability considerations into the structural design process would proceed more smoothly if the design engineers had some basic concrete "know-how". Unfortunately, concrete "know-how" typically does not form a part of today's young engineer's education and training. Neville [2] has observed that by "...not teaching something so ordinary, so simple as concrete (which could actually dirty your hands)...", the status of the teaching institution is actually enhanced. In many North American universities at the undergraduate level, the teaching of concrete fundamentals is typically combined in a "strength of materials" course along with many other types of materials. The total exposure to concrete is perhaps only several weeks. The student may subsequently be exposed to reinforced or prestressed concrete design studies, but design can be done with no real appreciation or understanding of the materials behavior. This is, of course, the underlying reason for this workshop. The real exposure to concrete appears to be at the graduate level. Unfortunately, the bright young engineers who go on to graduate school and specialize in concrete are very few with many of those remaining in academia rather than becoming designers.

It is often argued that there is not enough time in the curriculum to specialize in such a limited (and mundane) subject as concrete, yet there always appears to be time for other specialized subjects relating to our computerized (and perceived exciting) future. In a way, you can't blame the universities as they must offer a program that attracts students if they want to remain in business. On the other hand, concrete is the building material of choice in almost all regions of the world and certainly deserves more attention than it gets. Can we change the way universities establish their priorities and include a full 16 week course on concrete as the author had 42 years ago? Probably not in the short term, but we can at least begin trying.

Neville [2] also noted that the problem is not limited only to young engineers. More senior people may have basic "know-how" but typically are not in a position to mentor young engineers. They typically are in the office while the newer, younger engineers are in the field and have to learn basically by trial and error. Also, the more senior engineers generally have a certain unwillingness to accept newer (not new) methods and materials. The author dealt with a "senior" individual several years ago whose firm had just received a contract to produce and deliver over 165,000 m^3 of 80 N/mm^2 concrete. He insisted that he could produce this concrete using a cumulative weigh batcher and truck mixing as that is the way they always produced concrete in his area. Upon closer investigation, it was determined that his firm had never had never produced over 40 N/mm^2 concrete and his equipment and procedures were satisfactory for that quality of concrete. At the owners insistence, he converted to individual materials weigh hoppers, high shear stationary mixers, and improved quality control systems and now routinely produces 80 N/mm^2 concrete and wonders why he never did that before. Once he accepted a newer system to produce the concrete, he became "educated", in a small way, about the continuing improvements in concrete technology.

The problem of re-educating the more senior people is simpler than getting the basic education problem solved. Through the use of short courses (1 to 3 days) on specific subjects relating to the focus of this conference, senior engineers can rapidly be informed of the new technology. These courses should be taught by concrete technologists and be organized and administered by National concrete organizations. The venue should be such that it is easily accessible at limited costs.

TIMING

Consideration of the materials selection at the design stage must be given the same level of attention as given structural considerations. Common sense dictates that if a structural member must accommodate certain loads at given levels for the life of a structure, then the materials which comprise that member must maintain their integrity for the same time period. In many instances, however, the choices of ideal materials and techniques may clash with preconceived ideas on design and construction procedures employed by engineers.

This goes back to the problem of education of the design engineers and also the owners. If the owner can be convinced that he must involve a concrete technologist at the earliest stage of his planning, then there should be sufficient time to get the materials evaluated so a proper selection can be made. Similarly, newer or innovative methods of construction can be considered and included in the bid documents for the work.

As an example from the author's experience, the same job noted above initially planned to use all local materials to produce the 80 N/mm^2 concrete. By involving a concrete technologist in the planning stage, it was determined that the local cement had an equivalent alkali content greater than 1.2% and that all local aggregates ranged from slightly to highly reactive to alkalis.

The production of 80 N/mm^2 concrete also involved high cementing contents which put even more alkalis in contact with the aggregates. By initiating an alkali-aggregate reactivity evaluation program approximately 1-year before the award of contract, suitable materials for consideration by the contractor were identified. Upon award of the concrete, no delays in start-up were encountered because of materials selection. It was estimated that the cost of doing that preliminary work was only 0.03% of what the delay costs would have been.

CONSTRUCTABILITY

Many problems attributed to contractor deficiencies are caused by a lack of understanding of construction operations by the designer. These problems can be related to reinforcement details, section size and shape, embedments and penetrations, obstructions to casting or consolidation, and limited access for both construction and inspection. The designer may not be wholly at fault as the design may only reflect the Code requirements or the owner's wishes. It is the author's opinion that a constructability review by people sensitive to the durability issues be made for every major construction. While reviews may be common practice in certain industries, the question of whether durability aspects are considered in those reviews is not always answered.

Simple details, such as having a slight slope on sills to avoid the collection of water or salt on them, can help avoid a potential problem of concrete deterioration. Yet, the designer will probably include a horizontal surface as it is easier to consider and the constructor will prefer to build a horizontal section as it is easier to construct. Many other simple changes for durability enhancement, such as rounded corners rather than orthogonal corners, could probably be identified if durability of the concrete was considered at the early stages of the project.

Poor construction practices will always persist if proper quality control and quality assurance procedures are not enforced. Training aids for concrete craftsman have been developed by several organizations and have been extremely beneficial in providing guidance to the workers on what is the proper way to do concrete construction. This training must be provided to the workers at the start of the project and reinforced as the project progresses.

With respect to concrete durability, the assumption that a qualified constructor knows what he is doing can get a structure in trouble very quickly. A constructor with a good reputation may have known what "good practice" was on jobs that were done years ago, but personnel continually change in that industry and the people he has today may not know what was done before and will bring to the project their own experience which may not be good.

When someone says they have 25 years of experience in concrete construction, they never qualify it by saying this was good experience or bad experience. As an example, the author recently reviewed a proposal from a very reputable constructor to accept a construction practice which is universally rejected by most building codes and standard practices. In the proposal, documentation was provided which indicated that many years ago, the contractor had convinced an owner to accept the proposed practice based on limited testing by the contractor. This testing had mixed results and was subject to different interpretations. Its implementation was only to make the construction scenario a little simpler for the constructor. This deviation from accepted practice was also for construction in a certain environment

The change in practice impacted the potential for corrosion of prestressing strands. Based on this owner's agreement to the change, the next owner was told that the first owner agreed to this so he should also. And he did. The third owner was presented with two prior acceptances and he also agreed. And, it continued with this proposal despite changing structures, materials, and construction environments. When challenged on this, the constructor says "This is the way we've always done it." and that the accepted current practices are "too restrictive".

Who's at fault? Probably all parties, including the owners for not considering the ramifications (risk) of accepting such changes. If the owner had considered the durability aspects of his structure through the use of competent concrete technologists, he probably would never have accepted the change in practice. The need for good construction practices is essential in getting good durability performance. "This is the way we've always done it." is not always the best answer if that way is part of the problem. How can we ensure the constructor will use good durability related practices. It goes back again to education. A good start is to chose a reputable constructor and a proper specification that describes specifically what durability practices are to be implemented. That, by itself, may not be sufficient because there are no guarantees that the specification will be followed. As noted before, a proper QA/QC program is also required.

CONCLUDING REMARKS

The approach to integrating durability design into the structural design process will meet the least resistance if the requirements for durable concrete can be initially included in our national codes and standards. These requirements must be realistic and obtainable. The production of more durable concrete will lead to higher strength concretes. The codes and standards can then undergo future revisions to effectively use the additional strength.

Strong guidance on defining the environment in which the concrete will be used is necessary. This will require more work by the preparers of the codes and standards because current definitions are too general and only consider fairly simple scenarios. To this end, the concrete research community must take a very active role.

A service life requirement should be specified for every structure. Although there can be considerable uncertainty about what a proper service life should, a conservative estimate should always be selected. The incremental costs for improving the life of concrete from 50 to 100 years are infinitesimal compared to the costs of repairing or replacing a structure in the future. The service life selected should, however, be fairly realistic consistent with the planned obsolescence of the structure.

The performance criteria to be included in the codes and standards should focus on the transport properties of the concrete as it is the ingress of the environment into the concrete that is the principal cause of concrete degradation. The necessary criteria generally exist within concrete research community, but must be synthesized and formalized. Prediction of long-term concrete behavior involves the extrapolation of laboratory data, based on the assumption that long-term processes, which typically aren't known, will not invalidate the extrapolation. Additional work is still necessary. Because of this uncertainty, the performance criteria should be complimented by some prescriptive criteria that requires the designer to use minimum cement contents, maximum water/binder ratios, supplementary cementing materials, and chemical admixtures. The designer and constructor can not be expected to know what these should be. A strong position must be taken by the preparers of the codes and standards on this point. If we are serious about providing durable concrete, binder contents of less than 400 kg/m^3, water/ binder ratios more than 0.38, and prohibitions on the use of silica fume, fly ash or blast furnace slag are not acceptable for any concrete to be used in an outside environment. Some specifications for specific environmental conditions or classes of concrete have already begun incorporating minimum cement contents or maximum water/binder ratio requirements [3,4,5]. While their requirements are only approaching the suggested limits above, it indicates that this type of solution is recognized by the code writing people. It just needs to be expanded.

The evaluation of performance of the concrete should use National standard tests wherever possible. If there are no such tests, the concrete research community needs to establish and standardize such tests. This may require concessions by many researchers who believe that the method they developed is the only method of merit. Resolution of similar matters like this in other areas of concrete testing has been successfully done and can certainly be achieved for methods of evaluating transport phenomena.

Having done all of the above, there will still be resistance to the concept of providing high performance concrete on a routine basis. The cost of improved durability, from research through demonstration, will be compared with the costs of earlier, less ambitious efforts with the conclusion that the extra costs are not warranted. There is a great need in our industry for life-cycle cost analysis to highlight the need for more durable concrete. And finally, we must realize that durability is a complex topic and compliance with durability requirements may, in some instances, not be sufficient to ensure a durable structure. However, these requirements greatly increase the chances of success. We must also realize that the improvement of durability may not improve the performance of the concrete in other areas. For example, the use of fly ash may reduce early concrete strength development which may be needed for timely construction. There will always be exceptions, but these can be handled on an individual basis.

The above comments, suggestions and recommendations are the views of the author based on his extensive experience in dealing with concrete durability problems in many different regions of the world. Whether you agree with them or not, the need to integrate our knowledge on concrete durability into the design process is real and urgent and steps must be taken as soon as possible to accomplish this. Hopefully, the suggestions offered here are a beginning. The Shilstones [6] have often noted that the many building codes which include guidelines for producing durable concrete based on 28-day strength for selected water-cementitious ratios are the same codes that have guided the industry during the period in history where the durability of concrete has regressed. Either these requirements are not adequate, are not being followed, or are being negated by the use of improper construction practices. A closer look at all these aspects of the durability problem is needed. ***What we do know is that no concrete yet manufactured will continue to fulfill its role indefinitely.*** Hopefully, the convening of durability experts at workshops and international conferences of this type will take another step forward in producing the "eternal concrete".

REFERENCES

1. SOMERVILLE, G. The design life of concrete structures, The Structural Engineer, Vol. 64A, No. 2, February 1986, pp 60-71.

2. NEVILLE, A., Concrete in the Year 2000, Advances in Concrete Technology, 2nd Edition (Editor: V.M. Malhotra), Canada Centre for Mineral and Energy Technology, Ottawa, Ontario, Canada, 1994, pp 1-17.

3. CANADIAN STANDARD S474 (1994), Concrete Structures, Part IV of the Code for the Design, Construction, and Installation of Fixed Offshore Structures, Canadian Standards Association, Rexdale, Ontario, Canada, June 1994.

4. ACI 357R-84 (1997), Guide for the Design and Construction of Fixed Offshore Concrete Structures, Part 4, ACI Manual of Concrete Practice, 1998, American Concrete Institute, Farmington Hills, Michigan.

5. ACI 302.1R-89 (1996), Guide for Concrete Floor and Slab Construction, Part 2, ACI Manual of Concrete Practice, 1998, American Concrete Institute, Farmington Hills, Michigan.

6. SHILSTONE SR, J M, AND SHILSTONE JR, J M. Advances in Production Control of Mixtures for High Performance Concrete. Supplementary Paper, 2nd CANMET/ACI International Symposium on Advances in Concrete Technology, Las Vegas, Nevada, June 11-14, 1995 pp1-26

EXTENDING STRUCTURAL PERFORMANCE OF CONCRETE CONSTRUCTIONS

D Van Gemert

Catholic University Leuven

Belgium

ABSTRACT. Rehabilitation, upgrading and conservation of the vast stocks of old and degraded constructions allow financial and economical savings over new building. Extending the life span of concrete constructions is only possible if their structural performance is assured, and adapted to the expected future use of the buildings. The paper reviews recent evolutions in structural repair techniques, with emphasis on repair mortars, external steel and composite laminate strengthening, fire behavior and protection after repair. Assessment of repair performance, based on quality management of products, procedures and personnel will be presented.

Keywords: Rehabilitation, Strengthening, Performance, Quality management, Cement, Polymer, Fire protection, Personnel

Professor Dionys Van Gemert is professor of building materials science and renovation of constructions at the Department of Civil Engineering of K.U.Leuven. He is head of the Reyntjens Laboratory for Materials Testing. His research concerns repair and strengthening of constructions, deterioration and protection of building materials, concrete polymer composites.

INTRODUCTION

Public and private building owners nowadays realize the potential value of our vast stocks of old or degraded constructions as a means of providing modernized and appropriate accommodation for housing, utility and industry more quickly and at a lower cost than the alternative of new construction. Important financial savings can be obtained. The reasons for rehabilitation are numerous: the availability of existing obsolete, redundant, damaged or inappropriate constructions; the structural quality of these constructions; the shorter development period and economic advantages [1]. For rehabilitation of buildings additional advantages concerning permission issues, architecture, social aspects and others can be important.

The management of rehabilitation schemes is usually more complex than the management of new construction. The architect, the engineer and the contractor must be aware of the peculiarities and pitfalls of rehabilitation work. Without an appropriate initial building and structural survey, it is impossible to accurately assess the nature and extent of the work required and the likely cost involved. In assessing the necessary rehabilitation works, it is of vital importance to ensure that the building complies with current regulations, especially regarding the requirements for safety and serviceability of structures, as described e.g. in Eurocode 1 [2]. The 'design working life' is the assumed period for which a structure is to be used for its intended purposes with anticipated maintenance but without major repair works being necessary. Eurocode 1 indicates a design working life of 25 years for replaceable structural parts, 50 years for normal building structures and 100 years for monumental building structures, bridges and other civil engineering structures. Eurocode 1 does not explicitly cover assessment and re-design of existing structures, but it is quite clear that also for repair and rehabilitation the liability of designers and contractors extends far beyond the normal 10-years insurance period. This 10 years period particularly concerns the non-structural parts, e.g. concrete cover repairs at reinforcement corrosion and protective coatings.

Compared to this long design working life, experience with repair materials and repair techniques is still relatively short. Moreover, there is a strong and rapid evolution in the rehabilitation technologies, and in many cases structural rehabilitation techniques are based on the use of concrete-polymer composites, about which less experience is available. The long-term performance of rehabilitations is of prime importance for the involved design engineer, who is personally responsible for the safety of the building users, and that during a very long period. Long-term performance is an approximate synonym of durability: durability of a material or a building component is its ability to resist changes in its state or, in other words, of its properties. In rehabilitation, the durability problem concerns several aspects: residual performance or durability before and after rehabilitation; change of use; external reinforcement; assessing repair performance.

Extending structural performance deals with the prolongation of the working life of structures, as well as with the extension of their structural use and increase in mechanical or environmental influences on structures.

Of utmost importance in rehabilitation is real life experience, gained in case-studies. They fill the gap in theoretical knowledge about prediction of the service life or durability of materials and systems.

Evolutions in strengthening of reinforced and prestressed concrete constructions will be discussed, and criteria for the evaluation of the fire safety of these repairs are proposed. Fire hazards to externally bonded reinforcements are an obvious although not unsurmountable problem. Procedures for fire protection of the externally bonded reinforcements are presented. Besides strengthening, the load bearing capacity of concrete structures may be endangered by reinforcement corrosion. Evaluation criteria for repair mortars for concrete corrosion repair are given, and procedures for the performance assessment of structural repair materials and procedures are presented.

EVOLUTIONS IN EXTERNALLY BONDED PLATES

Plate Bonding Technique For Solving Concrete Damage

Concrete constructions can fail due to a variety of causes. Each of the damage causes, listed hereafter, covers a nearly unlimited number of specific situations, which have their own story, dramatic or amusing [3].

Calculation errors sometimes occur. If they only concern strength, they mostly remain undiscovered, because of the material safety and load factors taken into account at the design, and because of the ductility of the materials and the building structures. The situation is different when stiffness is involved. Once the total deflections exceed about 1/300 of the span, a psychological threshold is passed and people become uncomfortable in the building. In such cases the deflections must be eliminated by lifting of the concrete beam or plate. In general, due to plastic deformation of the concrete this will only be possible through cracking of the compressed zone of the concrete cross-section. As a consequence, the concrete element must be stiffened in the tension zone as well as in the compression zone [4].

Execution errors are more frequent than calculation errors. Externally bonded steel plates or fiber reinforced laminates provide a simple remedy if cracks arise in beams due to faulty overlaps of the reinforcements, if the continuity reinforcement rods in beams or plates are missing, if a supporting steel beam under a reinforced concrete plate is forgotten or if the connection between the steel beam and the concrete plates at both sides of the beam is badly designed or executed. Externally bonded reinforcements have a great flexibility to provide solutions for such damage cases.

Change of use of a building leads to other live or permanent loads on the structure, and even the structural geometry can be changed. External reinforcements can be applied to obtain considerable increase in bending and shear strength, where required.

Mechanical and other actions are a comprehensive cause of structural damage. Most striking are explosion damages and accidents [5]. Overloading is another action that may cause heavy damage to concrete structures. Environmental effects can destroy the concrete strength, e.g. freezing of balcony plates. Similar damage may be caused by fire. Creep of concrete may affect the deflections. Alkali aggregate reaction leads to destruction of concrete elements. Human mistakes often occur at the building site during construction: for the technical installations holes are drilled through columns, beams and plates with diamond cylindrical tools. If no precautions are taken, the risk of damage to the internal reinforcements is obvious. In all these cases repairs or protective treatments can be designed using epoxy bonded plates or fiber laminates [6].

Durability of External Reinforcement

Extending structural performance in time requires information about the durability of the bond against creep and environmental effects. Since the late sixties the technique is being used more and more, and its to day's popularity is the clearest proof of its durability.

However, extensive research has been done to study the long term behaviour of the glued composite: steel plate, epoxy adhesive or concrete. Pioneering work was done at EMPA-Zürich [7]. Corrosion and creep measurements were undertaken on a great number of long term exposed beams (1950 mm x 200 mm x 150 mm) strengthened with an externally bonded steel plate. This project started in 1977 and went on more than 10 years. After the exposure period no difference in static behaviour up to failure was observed. Different preparations of the bonding surfaces were compared, to check if rusting and debonding had occurred and, if so, how it could be prevented.

Additional tests were carried out in other countries [8]. Grit blasting proved to be the most appropriate preparation for both the steel and concrete bonding face. The epoxy adhesive, with and without primer, seems to have no significant influence on the corrosion behaviour. Creep of the adhesive layer is slightly higher than normal concrete creep. The creep difference increases with time although the variation is very small. It does not affect the mechanical behaviour of the steel-concrete composite element. Fatigue tests showed no fatigue failure, except at very high tensile stress variations ($\Delta\sigma = 310$ MPa) in the bonded plate. In that case plate separation from the concrete can occur, but the fatigue cracks can propagate in the concrete, as well as in the adhesive layer, depending on the specific system conditions [9].

Based on these positive experimental results, a durable external reinforcement will be obtained if the following conditions are fulfilled:

- relatively hydrophobic adhesives must be used, with a suitable rheology to ensure complete wetting of the steel surface;

- grit blasting is the minimum appropriate steel surface pre-treatment;

- stress levels in the anchoring zones must be limited, especially in fatigue loading;

- design engineers and contractors must be trained and familiar with the technique. Only certified adhesives should be used.

One of the methods for checking the durability of the glued connection under atmospheric conditions is the tear-off test on small cylinders, glued to the concrete surface. A great number of cylinders were glued on three repaired bridges over the Nete-Canal at Lier (B). Over a ten year period each time, six cylinders were tested both in the summer and winter period. After 10 years no decrease in strength could be observed [1].

The rheological behaviour of an epoxy bonded steel-concrete connection is being studied on a concrete plate reinforced with two epoxy bonded steel plates only. The loaded plate is shown in Figure 1.

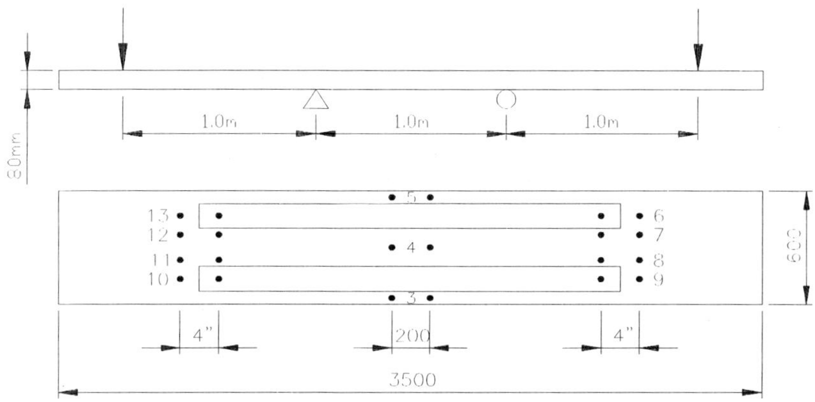

Figure 1 Externally reinforced concrete plate for outdoor exposure

The numbers indicate the different measuring bases. The loads, applied in a third-point bending test, were applied on 30-day-old concrete, whilst the glue had been cured for 7 days in laboratory conditions. The total load was 9.6 kN.

The creep test started on 05th November 1981. The rheological behaviour of the glue is evaluated by comparing the evolution of the strain on the steel-concrete bases with the evolution of the concrete-concrete strains. The evolution over 16 years is shown in Figure 2.

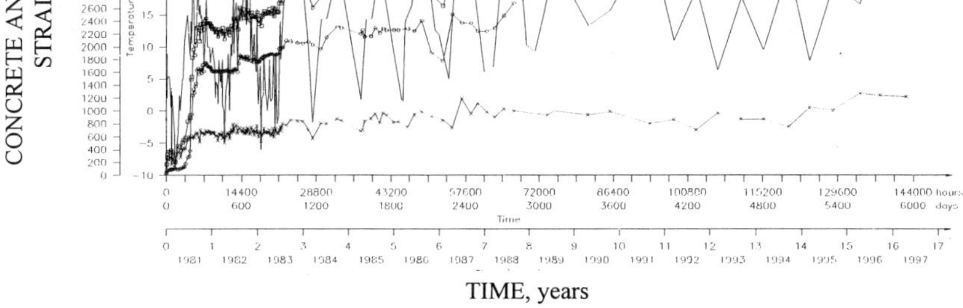

Figure 2 Evolution of temperature and strains in period 1981- 1997

Studying the temperature evolution on the same figure indicates that the first warm period had an important effect on the behaviour of the glue. An analogous variation has not been measured since. The differential creep seems to increase with time, although the variation is very small. On 12th August 1997 at 5 pm the plate structure collapsed, as shown in Figure 3.

Figure 3 Strengthened Plate, collapsed after 16 years of oudoor exposure

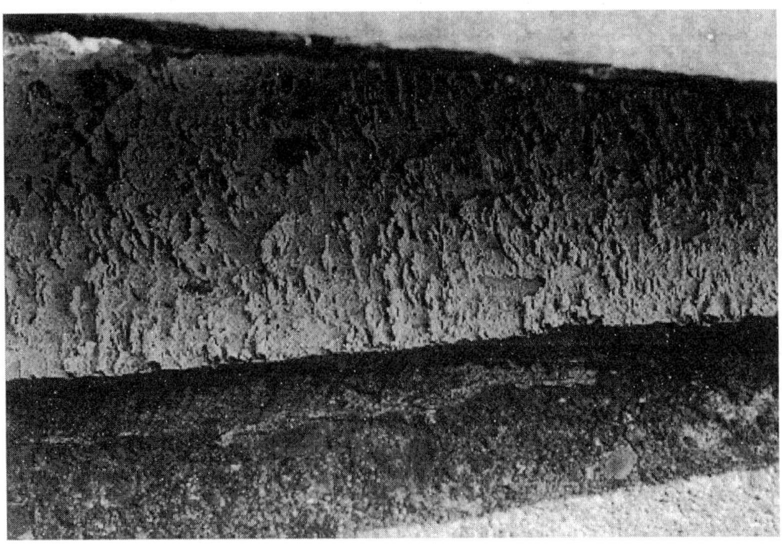

Figure 4 Stretched surface of glue layer

The epoxy bond between steel plates and concrete failed after 15 years and 9 months of climatic exposure. The collapse took place at 17h30, at which moment the ambient temperature was still 30 °C after a very hot day with maximum temperature of 37 °C. Intense vibrations had been caused during the day by rehabilitation works on the sewer system at a distance of 2 m from the plate support. The theoretical shear stress in the glue layer was about 0.12 N/mm², but the maximum shear stress at the plate ends could amount up to 1.5 N/mm². The shear span to depth ratio was 1000/80 = 12.5 and the unplated length 400 mm. The rupture of the epoxy glue was due to viscous flow of the epoxy. This can be derived from the surface aspect of the glue after rupture, as shown in Figure 4.

The surface shows successive zones, where the glue has stretched between the steel plate and the concrete. The collapse time, several hours after the hottest moment of the day, proves that the collapse happened through an accumulation of damage in the joint, amplified by viscous flow. The fracture also happened at the plate ends, which were most exposed to direct sun radiation and heating.

Carbon Fibre Reinforced Epoxy Laminates (CFRP)

Evolution in the technology of new materials, makes it possible to replace classic steel plates by new high-grade materials. This has led to the idea of replacing the steel plates by fibre reinforced composite sheets made of unidirectional, continuous fibres such as glass, carbon and polymers bonded together with a matrix such as epoxy resin. For strengthening concrete beams, carbon fibre reinforced epoxy laminates are the most appropriate. In the early stage these laminates had to be autoclaved, which was difficult to execute in practice. The laminates were delivered on roll, Figure 5.

Since the availability of so-called prepreg laminates and pure UD-sheets, Figure 6, these difficulties have disappeared. Prepreg epoxy laminates are preimpregnated with an epoxy resin which holds the fibres together. The prepreg sheets are very flexible and can be cut easily by means of scissors. At application the prepreg sheets are impregnated again with the right ratio of epoxy resin components and the chemical reaction starts. When the first layer has hardened enough, the second layer can be applied in the same manner. Up to 10 layers can be applied. UD-sheets are completely impregnated on the site.

These prepreg and UD-sheets were first developed and produced in Japan. In Belgium the first application of on site hardening CFRP-laminates took place at the beginning of 1996 [6].

The CFRP laminates have many advantages over classic steel plates. The mechanical properties are superior. The tensile strength of the prepreg sheets is 5 to 10 times higher than that of steel, whereas the modulus of elasticity is comparable to steel. Some carbon fibre laminates reach a modulus of elasticity up to 650000 MPa. These good mechanical characteristics allow a smaller cross-section of reinforcement. Prepreg sheets are also easier to process. Since the CFRP laminates are much lighter - the density is about 3 times lower than the density of steel - the sheets can be placed with less manpower.

The prepreg sheets are available on roll which means that they are available in any length, whereas the steel plates are limited in practice to 6 metres.

Carbon fibres are very corrosion resistant. An expensive surface treatment, as for steel, is not necessary. The CFRP laminates are extremely useful in very corrosive atmospheres, such as marine and aggressive chemical atmospheres. The very thin (0.2 - 0.4 mm) laminates are very easy to apply and can be applied in cross directions without any difficulty. This is a great advantage for strengthening of two-way slabs, especially because now orthotropic behavior needs no longer be taken into account.

There are some disadvantages too. First, it is a brittle material; the carbon fibre behaves in a linear elastic way without a plastic phase up until rupture. The rupture occurs without preceding plastic deformation. The anchorage can give problems too. Due to the high stresses, the CFRP-laminates will peel off easily. The carbon fibers are longitudinally oriented and the good mechanical properties are only valid in that direction. Without special precautions it is impossible to drill a dowel hole. The fibers would be cut and no longer able to transfer forces. Special stirrup layout can offer good solutions to avoid plate end shear cracking or plate separation [11,12].

Usually it is impossible to exploit the high strength quality of the carbon fibre laminates. High stresses can only be developed at large strains and that may cause excessive stresses and strains in the concrete section. Prestressing of the laminates offers a theoretical solution to this problem, but in practice prestressing the laminates on the renovation site is mostly very difficult, not to say impossible. Finally the price for CFRP-laminates is rather high. The material is much more expensive than steel but the processing cost is much lower.

Further research should be dedicated to the fatigue resistance of the composite system: concrete/glue/laminate. Bond and crack growth, as well as delamination within the laminate must be studied. In this respect, the effect of bolts and anchorage lay-out on bond, crack growth and delamination are important aspects.

In those applications where expansion is not a problem, simple wrapping may satisfy to strengthen the elements, e.g. wrapping of columns for strengthening bridges in seismic areas. Carbon fibres may be replaced by high performance glass or other fibres. Special applications of CFRP laminates are envisaged, e.g. application for wrapping concrete columns, damaged or endangered by alkali-aggregate reaction. The column expansion of concrete damaged by alkali-aggregate reaction can be to 10 %.

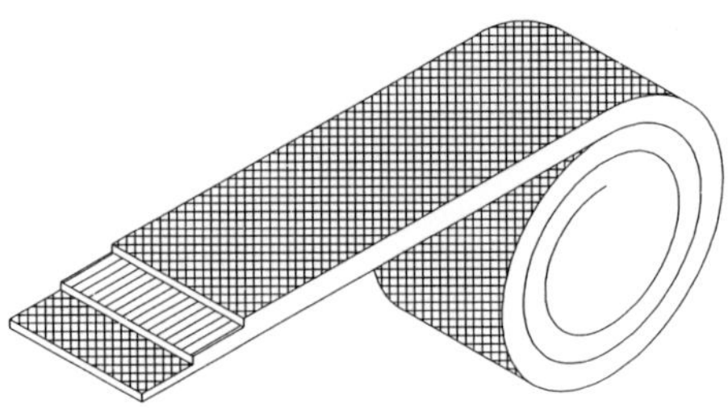

Figure 5 CFRP-sheet on roll

The wrapping laminates would not be able to allow for this expansion. Therefore research should concentrate on the production of a hybrid laminate, consisting of a 3D-sandwich fabric, as shown in Figure 7.

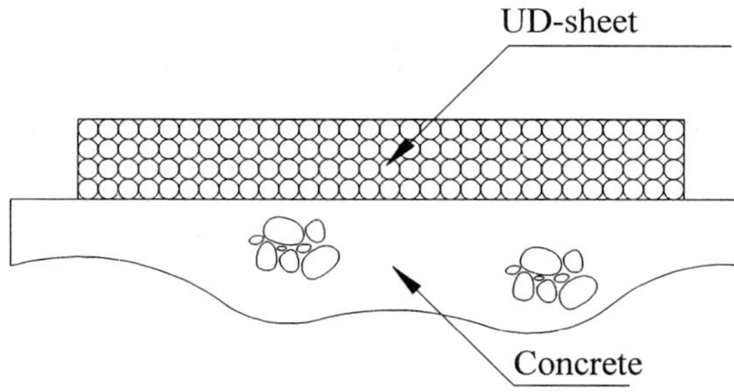

Figure 6 UD-sheet

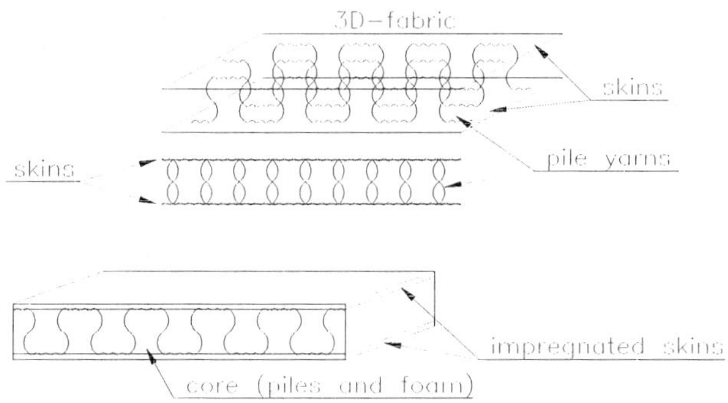

Figure 7 3D sandwich panel

Such a 3D fabric can be made to harden on the site, after application, whereby the piles can be made to stretch during hardening. In that way a stiff and strong laminate that supports the damaged element is obtained; if there is excessive expansion of the concrete section, the inner laminate skin may break, but the remaining fabric will be able to carry all loads. The core, consisting of foam and piles, must be able to absorb the swelling without damaging the outer skin of the fabric, Figure 8.

Fire Protection of Externally Bonded Reinforcements

Fire is likely to damage externally bonded reinforcements. The cold hardening glues have a glass transition temperature of about 60 to 90 °C. During a fire that temperature is easily and quickly attained if the reinforcements are exposed directly to the fire. However, the external reinforcement is usually applied to improve the degree of safety, and the total action of permanent and variable loads can be carried by the concrete structure alone, be it without any safety margin. During a fire a very low safety margin can be allowed, e.g. 1.05 to the ultimate load. If this safety is still available in the structure without external reinforcement, than one can decide to abandon a special fire protection. In most practical cases enough safety in the non-strengthened situation is still available, so that fire protection is usually unnecessary. Moreover, the investment cost for a fire protection in general is higher than the repair or strengthening cost, which makes it mostly uneconomical, compared to the risk of fire and the risk for people in the case of fire.

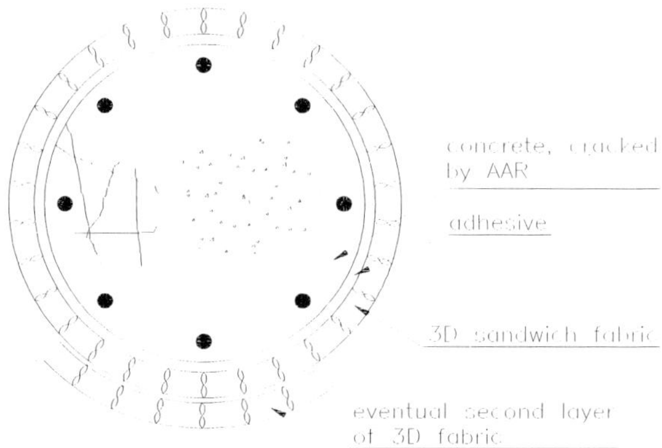

Figure 8 Strengthening of concrete column damaged by alkali aggregate reaction

If for special reasons a fire protection is required as yet, special protective coverings are available [13]. Epoxy bonded external reinforcements can be protected by means of insulating boxes, made out of fiber reinforced silicates.

REPAIR MORTARS FOR STRUCTURAL REHABILITATION

Classification of Repair Mortars

Repair mortars for concrete are subdivided according to their use and according to the binder material.

It is necessary to distinguish between mortars that should contribute to strength and stiffness of the structure, and those who are not expected to do so. Additional requirements can be imposed regarding mechanical and physico-chemical resistance of the materials, e.g. abrasion, carbonation, diffusion.

The German regulations [14] define 4 classes of mortars:

M1: mortar appropriate to re-profile damaged zones; re-establishment of alkaline environment around reinforcing steel; sufficient bond to substrate; convenient substrate for additional protective layers.

M2: requirements for M1, plus carbonation and chloride ingress resistance; application possible under dynamic loads.

M3: repair of stressed cross-sections; requirements for
- strength and deformation, shrinkage and creep
- adhesion to reinforcing steel
- bond to substrate
- durability at elevated temperatures
- durability under sustained moist conditions.

M4: requirements for class M1, plus strength and abrasion resistance.

It is clear that each class contains mortars with different binder type, and one mortar can belong to several classes. On the basis of binder type, 3 groups of mortar are distinguished:

1. mineral mortars on cement basis, eventually with admixtures; concrete and spray concrete;
2. polymer modified mortars (PCC); spray mortar (SPCC);
3. polymer mortars.

The difference between repair mortar and concrete is the aggregate size: up to 4 mm for mortar, bigger for concrete. The maximum grain size should not exceed 1/3 of the layer thickness. Cement based mortars are applied wet in wet, and curing is necessary. A good bond remains problematic, and thin repair layers (< 20 mm) are inappropriate.

Polymer modified mortars contain a polymer resin component, added in the mix as a dispersion in water or as a dispersible powder. Recently, active functional groups such as silanes are grafted on the polymer molecular chain. They act as bond promotors because they adhere chemically to the polymer and to the aggregate surface. This improves the life time and performance under moisture conditions. Structure and composition of these mortars are close to those of the concrete substrate, avoiding major compatibility problems.

Polymer mortars are completely different from concrete, and their use is only justified if special performances are required:

1. very high chemical resistance and bond
2. fast hardening
3. curing of mineral or polymer modified mortars not possible
4. thin layers.

Typical short-term properties of the three types of mortar are given in Table 1.

More extensive discussion of repair mortars is available in literature [15, 16]. An important property concerning compatibility is shrinkage: plastic and drying shrinkage.

Shrinkage is a complex function of a multitude of factors. A non exhaustive list is given in Table 2.

Because of all these influencing factors a 0 % drying shrinkage is the target to obtain a good mortar performance, Figure 9 [16].

It is quite difficult to realise this goal, especially because the mortar passes different consistency phases during its hardening process, at which firstly expansion and afterwards shrinkage occurs. Depending on circumstances, e.g. restrained or free movements, this may result in a greater or smaller final shrinkage.

Table 1 Short term properties of repair mortars

PROPERTY	POLYMER MORTAR	POLYMER-CEMENT MORTAR	CEMENT MORTAR
	PC	PCC	CC
Compressive strength (N/mm^2)	50 - 100	30 - 60	20 - 50
Tensile strength (N/mm^2)	10 - 15	5 - 10	2 - 5
E-modulus (kN/mm^2)	10 - 20	15 - 25	20 - 30
Thermal expansion (per °C)	$20 - 30 \times 10^{-6}$	$10 - 20 \times 10^{-6}$	10×10^{-6}
Water absorption (w %)	1 - 2	0.1 - 0.5	5 - 10
Max. service temp. (°C)	40 - 80	100 - 300	> 300

Table 2 Factors inflencing the drying shrinkageof cement based mortars

MATERIAL PARAMETERS	ENVIRONMENT
- cement content	- relative air humidity
- cement type	- drying rate
- water content	- drying time
- aggregate ratio	
- admixtures	
- additives	
- porosity-compaction	
- age	
- volume/surface ratio	

Besides dimensional compatibility one has to take care about chemical compatibility as well. This concerns alkali-, C_3A-, Cl^--content etc. Electro-chemical compatibility concerns resistivity, pH. A concrete substrate containing reactive aggregates will be repaired with a low-alkali repair mortar. To avoid corrosion a repair mortar will be chose that does not make a macrocell with the substrate. Coating of a potential anodic zone will drastically change the cathode/anode surface ratio, causing accelerated corrosion. Covering the cathodic zone, where moisture and oxygen ingress must be prevented, will be th right action here.

The pull-off test is generally used to characterize the mechanical quality of the concrete surface and of the adhesion of repair materials on the concrete substrate. Not only the strength value is important, but also the type of rupture: cohesive in mortar or substrati; adhesion in interfac; mixed adhesion-cohesion. The square optimization method, developed by S. Pareek [19], is an elegant evaluation method for adhesion data.

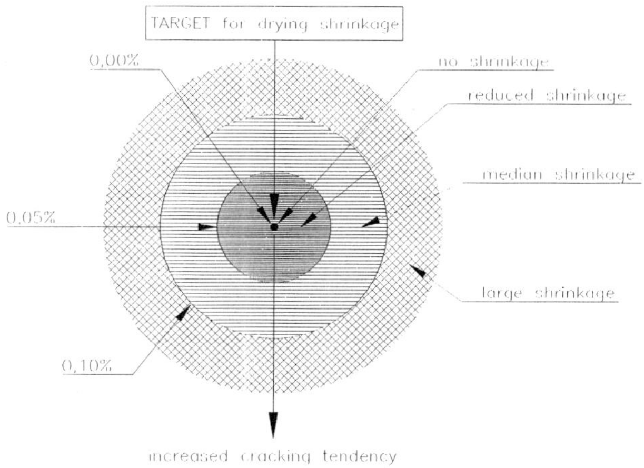

Figure 9 Targets for drying shrinkage

ASSESSMENT OF REPAIR PERFORMANCE

Polymer-modified concrete and mortars play a very important role in repair, strengthening and retrofitting of concrete structures. A lot of research has already been done in laboratories as well as by producers and ministries to set up codes of good practice to ensure durable repairs and treatments [14,17,18]. Due to a lack of mathematical or theoretical models for performance and durability predictions, most codes are based on a vast experimental testing and approval program, in which the following subdivisions are made:

- strict chemical and physical identification of material components
- mechanical and physical performance testing and approval
- qualification of personnel
- reception of materials on building site
- quality testing of executed rehabilitation work on site.

Most of these codes and procedures are in continuous evolution, and national committees are involved in preparing appropriate testing programs for special products or systems. An example are the technical agreement procedures, developed by UBAtc, the Belgian union for technical agreement in construction, who is a member of the European union UEAtc. The procedures are developed by a working group CEP-LIN-MET, in which public and private users, producers and research and control laboratories are represented. Sis sub-groups have been established, dealing with injections (SG1), structural strengthening (SG2), repair and protection of concrete and masonry (SG3), thin repair systems (SG4), special intervention techniques (SG5), quality assessment, diagnosis techniques (SG6).

These sub-groups prepare guidelines (Gxxx) for technical agreement, to be approved by the working group. Up to now about 20 guidelines have been prepared. The agreement procedure for a specific product is coordinated by executive boards. Four executive boards are already active: for coatings, thin repairs and repair mortars; for injection products; for strengthening by plate bonding; for special techniques. All the guidelines have the same structure:

- clear determination of the subject of the guideline
- terminology
- application field of the material subject to the guideline
- description of the material and application rules
- quality requirements
- description of the testing procedures
- presentation of materials for agreement testing
- quality control at the production site
- content of the agreement
- course of the agreement procedure.

The activities of the working group are in line with the European standardization efforts, coordinated by CEN. However, European standards are not yet fully available. The first standards are being published [20], and more standards are in preparation.

Besides this quality assessment procedures for materials and systems, efforts are being done to improve the quality of the personnel as well. In Belgium the certification of personnel is coordinated by the Ministries of Infrastructure and Environment, Division Control of Concrete Structures. The technicians have to prove their skill in applying the specific repair or strengthening procedure. The tests that have to be taken by the technicians are described in the ministerial document V576-B5. The examination is taken in a certified laboratory, under the supervision of a public servant.

CONCLUSIONS

Extending structural performance of concrete structures is a complex and comprehensive problem, involving technical, economical and managerial aspects. New techniques have been developed for strengthening of concrete structures by means of external steel plates or fibre-reinforced laminates. Repair mortars are now reliable components in structural rehabilitation and strengthening systems. The durability of the rehabilitation measures is being secured by a profound quality control of materials, and by an appropriate certification of the technical, executive personnel. Experience gained in case-studies and laboratory research programmes provides the calibration data and other information required to adapt and improve the codes and standards.

REFERENCES

1. VAN GEMERT D. Durability of rehabilitation, Proceedings Concrete in the service of mankind, Dundee 1996, Concrete Repair, Rehabilitation and Protection, Ed. R.K. Dhir, E & FN Spon 1996, pp. 549-558

2. CEN-ENV 1991:Eurocode 1 - Basis for design and actions on structures, CEN-Brussels, 1994

3. VAN GEMERT D. Design, application and durability of plate bonding technique, Proceedings Concrete in the service of mankind, Dundee 1996, Concrete Repair, Rehabilitation and Protection, Ed. R.K. Dhir, E & FN Spon 1996, pp. 559-570

4. VAN GEMERT D. Repairing of concrete structures by externally bonded steel plates, ICP/RILEM/IBK International Symposium 1981, Prague, Plactics in Material and Structural Engineering, pp. 519-526

5. VAN GEMERT D. Evolutions in design and application of the plate bonding technique, Symposium on Structural Repairs/Strengthening by the Plate Bonding Technique, Sheffield, G.B., September 6-7, 1990

6. VAN GEMERT D. Evolutions nouvelles en matière d'armatures collées pour le renforcement du bétonProceedings Conference 'Armatures défaillantes! Que faire?', Liège 20.03.96, 10p.

7. LADNER, M. Belastungsversuche an 12 fünfjärigen Stahlbetonbalken mit aufgeklebter Bewehrung, EMPA Bericht nr 37'846/4, 1983, Zürich, Switzerland

8. HOBBS B., JONES R., SWAMY N., ROBERTS M. Accelerated tests to assess the durability of bonded plate systems, Int. Seminar Structural Repairs/Strengthening by the plate bonding technique, Sheffield, 1990

9. HANKERS CH. Strengthening of structural concrete members by bonded steel plates. Fatigue behavior and fatigue design, Int. Seminar Structural Repairs/Strengthening by the plate bonding technique, Sheffield, 1990

10. VAN GEMERT D. Special design aspects of adhesive bonding plates, ACI SP 165, Ed. N. Swamy and R. Gaul, 1996, pp. 25-42

11. TÄLJSTEN B. Plate bonding. Strengthening of existing concrete structures with epoxy bonded plates of steel or fibre reinforced plastics, Doctoral Thesis, Luleå University of Technology, Sweden, 1994

12. VAN GEMERT D., AHMED O., BROSENS K. Anchoring of externally bonded CFRP reinforcement, Int. Congress Creating with Concrete, 1999

13. ETERNIT-PROMAT Fire prevention in constructions. Manual 91 (in Dutch), Eternit Group 1991

14. ZTV-SIB 90 Zusätzliche Technische Vertragsbedingungen und Richtlinien für Schutz und Instandsetzung von Betonbauteilen, Verkehrsblatt-Verlag, Dortmund 1990

15. VAN GEMERT D. Repair of concrete surfaces (in Dutch), Innovation in construction, K.VIV-TI Antwerp, 1996, pp. 167-178

16. EMMONS P., MCDONALD J., VAYSBURD A. Some compatibility problems in repair of concrete structures - a fresh look, Materials Science and Restoration, vol 1, pp. 836-857, Expert-Verlag 1992

17. OHAMA Y. Handbook of polymer-modified concrete and mortars, Noyes Publications, New Jersey, USA, 1995

18. IBAC-Aachen Schutz und Instandsetzung von Betonbauteilen. Die Neue DafStb-Richtlinie, RWTH Aachen, 04.10.1989

19. PAREEK S. Improvement in adhesion of polymeric repair and finish materials for reinforced concrete structures, PhD thesis, Nihon University, Koriyama, 1993

20. CEN EN1504 Products and systems for the protection and repair of concrete structures - Definitions, requirements, quality control and evaluation of conformity. Part 9, July 1997. Part 1, January 1998. Parts 2 - 8 in preparation.

WASTES IN CONCRETE: CONVERTING LIABILITIES INTO ASSETS

A Samarin

University of Wollongong

Australia

ABSTRACT. The paper presents a general overview of the current trends in the use of by-products or wastes in various types of concrete. A potential of using processed and unprocessed industrial by-products and domestic wastes as ingredients in concrete is evaluated. In particular, the feasibility of manufacturing hydraulic cements, non-hydraulic cements, cement *extenders*, admixtures and additives in concrete, unprocessed, processed and manufactured aggregate in concrete, and the use of processed or unprocessed by-products and wastes as concrete reinforcement is assessed. An attempt is made to predict possible future developments in the utilisation of industrial by-products as raw materials in cement and concrete due to ever increasing cost of waste disposal and due to the stricter environmental regulations. The impact of Kyoto Protocol, stipulating new levels of abatement of greenhouse gas emissions by the year 2012, on utilisation of cement *extenders* in concrete, and on the use of wastes as raw materials and fuel supplements in cement manufacturing, as well as the impact of the new technological developments on the use of wastes in concrete is evaluated.

Keywords: Broad definition of concrete, Industrial and domestic wastes, Processed and unprocessed wastes, Wastes as binders, Wastes as aggregates, New applications for wastes, Effects on the greenhouse gas emissions, Kyoto protocol, Economics of wastes utilisation, Environmental pollution, Sustainable development.

Dr Aleksander Samarin is a Private Consultant and Adviser in the Sustainable Development of Technological Sciences and Engineering and a Professorial Fellow at the School of Civil, Mining and Environmental Engineering, University of Wollongong, Australia. He is a Fellow of the Australian Academy of Technological Sciences and Engineering and a Vice-Chairman of the NSW Division of this Academy. Professor Samarin conducted R&D work into building and construction materials, energy and environment protection in many parts of the world, published widely, receiving several awards for his contribution to science and technology, and has taken a number of patents.

INTRODUCTION

In a very broad sense concrete can be defined as an artificial building and construction material consisting of relatively inert aggregate, held together with either inorganic or organic binder, or even by a hybrid of these two types.

Aggregate can also be either inorganic or organic in nature, it can be naturally occurring (processed or unprocessed), or manufactured. The reactivity of aggregate can also vary. Depending on the nature of reactivity of aggregate the effects can range from beneficial to highly destructive to concrete strength, dimensional stability and durability.

Industrial or domestic wastes, processed or unprocessed, can become cost-effective ingredients of concrete, defined in this way.

The processed or unprocessed industrial by-products or wastes can be used as raw materials in cement manufacturing, as components of concrete binders, or even as binders in their own right. They can also be used as aggregates, as a portion of aggregate, or as ingredients in manufactured aggregates.

Some wastes can be used as chemical admixtures and additives, which can alter and enhance selected, desirable properties of fresh or hardened concrete.

The successful use of industrial by-products or wastes in concrete depends on a number of factors. Of these, an application and nature of the end use of building and construction materials will dictate required properties of concrete, and thus will become a pre-requisite in deciding on whether or not the waste can be ultimately incorporated into this concrete. Desirable properties of concretes used for structural applications, or for block making, or in pavement construction, or for decorative use, etc., would obviously differ quite considerably from one another.

However, the economical factors would ultimately determine, if potentially beneficial waste can be used as an ingredient in concrete. These economical factors can differ considerably from one country to another, and particularly between developed and developing countries. They may also vary considerably for different locations in a given country, and are generally influenced by the cost of waste disposal, by the cost of transportation of waste to a manufacturing site and by the existing environmental regulations.

We can define waste as not readily avoidable by-product, for which there is no economical demand and for which disposal is required [1].

The first rule of waste management is to prevent its production in the first place, if it is at all possible. If not, we must attempt to generate an economical demand for the waste, for example by creating beneficial conditions for its recycling. Utilisation of waste, resulting in a cost-benefit to the end user is one way of creating a demand for a by-product. The benefits may include the elimination of quite tangible expenditure, such as cost of disposal, as well as some less corporeal advantages, such as abatement of pollution and better control of environmental contamination.

RECYCLED CONCRETE

Fundamental Requirements

After the second world war, when many European cities were reduced to rubble and there was a general shortage of cement and aggregate manufacturing facilities, the use of recycled concrete in building and road construction industries was reasonably wide spread.

The main problem with the use of recycled concrete, which existed in the late 1940-s, and which is still the principal source of annoyance in the present day use of recycled materials, is the difficulty of separating useful components of rubble from the contaminants, such as timber, glass, plastic, etc., which are often harmful to the properties of the end product [2].

The cost of "purification" of recycled rubble versus a tolerance to the impurities of concrete made with the recycled materials, generally determines in the end its potential suitability of use.

Rejuvenated Cement

Most of the current effort in recycling concrete is directed towards the re-use of coarse aggregate. The fine fraction of rubble, obtained from a hydraulic cement concrete, contains cement grains, which almost without exception are only partially hydrated. These cement particles can be "rejuvenated" by means of mechano-chemical activation and thus become either *cement extenders*, or even binders in their own right [3].

Recycled Aggregate in Road Construction

The most widely used form of concrete recycling is possibly in road construction. Bituminous concrete is particularly tolerant to the inclusions of all the components of old pavements, the main potential problem being badly oxidised bitumen. Addition of sufficient amount of fresh binder can overcome the problem. Special machinery has been developed for recycling of flexible pavements, mostly in the developed countries, although recycling of bituminous concrete is also practiced in the developing countries [4].

In my personal opinion *hybrid* or semi-rigid /semi-flexible pavements are a particularly good type of construction made from a recycled aggregate concrete [5].

Rigid pavements made with hydraulic cement concretes can be unreinforced, reinforced, or post-tensioned.

Failure of a rigid pavement can be either structural, or due to material deterioration in concrete. Structural failure often results from the settlement of a base course, from an inadequate provision of thermal expansion /contraction joints, or simply due to the traffic fatigue. The integrity of recycled concrete components in these cases is usually quite good.

When failure is due to the chemical or physical distress of concrete, such as alkali-aggregate reaction, frost damage, etc., the suitability of recycled aggregate for the new constructions work must be circumspect.

There has been a significant progress in the technological development of recycling methods of rigid pavements during the past decade [6], [7].

Generally the recycling work, using specially designed machinery, is done *in situ*. The old pavement is broken up, the rubble is crushed and screened. If the pavement was reinforced, steel rods are removed by electro-magnets or by mechanical means. The suitability if recycled material is assessed in a field laboratory, allowing design and construction of the new pavement, with the appropriate addition of new raw materials, to proceed without delay.

Recycled Aggregate in Building and Construction

Use of recycled aggregate in non-structural, non load-bearing concrete presents similar problems to those of aggregate used in pavements.

Requirements for structural concrete can vary significantly, depending on the type of structure, and within that structure they do change from one structural member to another. Similar structures in different environment conditions would behave in a dissimilar way, and hence would have specific provisions for the properties and the composition of concrete [8].

An impressive volume of research and development work is conducted in many parts of the world in this area, and there is no lack of published data on recycled aggregate in technical books, journals and conference proceedings [9], [10], [11], [12], [13], [14], [15], [16], [17].

Structural Standards do differ from one country to another, as they reflect environmental and economical conditions for the distinctive and sometimes unique parts of the world.

Thus, we can not expect to end up with universal standards or manuals for the use of recycled aggregate in structural concrete. Each specific application would have its own solution.

AGGREGATE DERIVED FROM WASTE

Unprocessed Waste as Aggregate in Concrete

A wide range of unprocessed waste, with variable degrees of success has been used as aggregate in concrete.

Probably the most widely researched is the use of blast furnace slag aggregate. The technology is well known and field results over thirty or even forty years, particularly in pavements, are available. Properties of blast furnace slag do vary considerably, and with the optimisation of smelting technology, have undergone considerable changes in the past thirty years or so. Australian blast furnace slags used to be glassy and dense, but with the use of fluxes, such as serpentine, to improve the furnace efficiency, became porous and even at times dendritic. However, most blast furnace slag aggregates are compatible with hydraulic cements, although free lime component of slag may cause dimensional instability in concrete, if slag is not properly weathered before use. The R&D work is continuing, and up to date reports are available [18], [19], [21].

Less widely used is steel slag, although it was reported that several steel slag aggregates performed well in Portland cement concrete [20], [21]. More R&D work, particularly into long term durability and also in the use of steel slag aggregate in semi-rigid pavements is required, in my opinion.

Some preliminary assessment of the behaviour of ferro-chromium and silica-manganese slags in concrete was also carried out [21].

Many building and industrial wastes can be used in special application concretes. Demolition waste, provided it is separated into the ingredients of similar properties can be used successfully in most non-structural and even in some structural concretes [22], [23]. The process of waste segregation, however, may prove difficult and therefore expensive [24], [25], [2].

The use of crushed brick or a mixture of crushed brick and of Portland cement concrete rubble as an aggregate in structural concrete was reported to be promising [26], [27]. Long term properties, which differ from conventional concretes, should be accounted for in the design. Modulus of elasticity of brick aggregate concrete may also differ from the Structural Code values, for a particular grade of concrete.

Crushed glass, from wine, beer or soft drink bottles, can be used as fine aggregate in concrete [28]. It is important however, to have sufficient content of either ultra fine glass particles, or of fly ash or other good quality pozzolans, to prevent alkali-aggregate expansion in larger glass particles.

Lightweight concretes can be produced with unprocessed waste materials, such as expanded polystyrene granules [29], [30]. There are several patented lightweight construction systems, incorporating expanded polystyrene concrete panels, which are now more than ten years old. A classical example is *Stracke Ges.m.b.H.* system in Vienna, Austria. Using special concrete mixer, lightweight Portland cement concrete with polyester aggregate of between 300 and 350 kg/m^3 is formed into blocks or panels, which can be sawn, drilled and nailed. These are then used as a permanent formwork in buildings, providing excellent thermal and acoustic properties to a reinforced structure, cast with high workability concrete.

Cork granules can be used in some of the above applications in place of expanded polystyrene chips [31].

Sawdust was used in different parts of the world, as aggregate for concrete [32]. Again, number of patents have been taken in these countries. Most describe the type of sawdust used, some special additives, preventing adverse reactions between the wood particles and Portland cement paste, mix proportions, and also mixing and curing techniques.

Typical sawdust concrete has a design strength of 15 MPa, density of approximately 1000 kg/m^3 and in some cases it can be sawn, milled and nailed. However, long term durability of this material, particularly in a wet-dry environment, needs additional investigation in my opinion.

An attempt to use shredded rubber as an aggregate in Portland cement concrete was generally not very successful [33]. Considerable reduction in strength was reported, even at relatively low rates of replacement of conventional aggregate with rubber. In my own, unpublished work it was found, that spherical rubber particles mixed into concrete, had an identical effect to the air voids of similar dimensions and total volume on the concrete properties. Use of rubber aggregate in semi-rigid concrete, however, can be quite promising.

There is an extremely wide range of different unprocessed wastes, currently being evaluated as potential aggregate in concrete. These include cane bagasse, wood ash [34], china clay waste, slate processing waste [35], paper waste, etc. These may be suitable for either temporary or low cost housing, particularly in the developing countries [36], [37], [38], [39].

Concrete Aggregate Manufactured from Waste

Work on the development of a process for conversion of pulverised fuel ash into a sintered lightweight aggregate - *Lytag* - for concrete, was carried out at the Building Research Establishment in Garston, U.K. since the early 1950-s, at the request of Central Electricity Generating Board. The R&D work was completed by the late 1950-s.

In the Lytag process, fly ash with a carbon content of between 7 and 10 percent is fed onto pan pelletisers, where the rotation action converts damp ash into spheroidal pellets. The pellets are then passed to the sinter strand, where they are fused at temperature of approximately 1300° C. Carbon in fly ash is ignited, producing gas and bloating the pellets, which, upon cooling, solidify to strong and lightweight particles. The first production plant of some 100,000 t/a was constructed in 1961 at Northfleet, U.K. [40].

Using Lytag, concrete with an air dry density of about 1,800 kg/m^3 can be obtained, with a maximum ultimate compressive strength of up to 70 MPa in individual, cube specimens. Thus, theoretically, characteristic strength of up to 50 MPa is achievable with special design techniques. In practice, it seldom exceeds 40 MPa. Long term behaviour and modulus of elasticity, however, can vary from the Structural Code values, for a given grade of concrete.

Lytag aggregate has been used also as a filtration material. There were several reports describing the manufacturing process and markets for Lytag in some detail [41].

There were number of attempts in the late 1980-s to convert household and industrial garbage into concrete aggregate. Commercial success of this concept depends *inter alia* on the price differential between the cost of aggregate production and the cost of garbage disposal.

Australian company, *Neutralysis* spent twelve million Aust. dollars over a period of six years developing the technology of manufacturing concrete aggregate from a blend of household and industrial waste. A pilot plant near Brisbane, Queensland, was built in 1991 [42].

The process starts with primary and secondary pulverisation of garbage blend accompanied by magnetic separation of scrap steel for recycling.

The garbage then enters the processing stage where it is mixed with the combustible liquid industrial waste and clay, extruded into pellets and dried. The pellets are fired in three successive kiln stages using excess energy given off during the process. In the final vetrification stage all the organic harmful materials are destroyed and inorganic hazardous materials are encapsulated. The aggregate was of the quality suitable for non-structural concrete, in concrete block manufacturing and as a road base material. The surplus energy can be used to generate electricity, which is then sold into the power grid.

Due to the relatively low cost of garbage disposal in Australia, the technology was sold to a sister company in the United States, where the waste disposal costs were considerably higher.

Another process, which converts the residue from incinerated toxic waste into pellets, was developed by a Singapore company, *Miltox.* Miltox, a joint venture between Singaporean, Malaysian and Australian interests built a pilot plant with an output of approximately 400 kilograms an hour in 1994. It seems, that long term research into the durability and environmental effects of the product is required, before it can be safely used in building and construction materials.

POZZOLANS AND OTHER PORTLAND CEMENT EXTENDERS

Applications in Concrete Technology

The use of fly ash, ground granulated blast furnace slag, silica fume, rice husks ash and other industrial or natural mineral by-products as extenders of Portland cement is very thoroughly researched and thousands, if not tens of thousand of papers, reports and books have been written on the subject. Wide spread industrial use of these materials have been almost universal for several decades. Number of International Conferences on the use of Fly Ash, Silica Fume, Slag and other Mineral By-products in Concrete were organised by ACI/ CANMET and were held in 1983 (Canada), 1986 (Spain), 1989 (Norway), 1992 (Turkey), and 1995 (USA), and the Proceedings were published by the American Concrete Institute as the Special Publication (SP) series. Many other conference proceedings, including conferences at Dundee, contain valuable information on the use of these materials. To go into the detailed analysis of all the published data is outside the scope of this paper.

There are however several very important points, arising from my own R&D work with pozzolanic materials, which I would like to emphasise. When fly ash was first used in Australia as a Portland cement extender in the 1960-s, I was often asked by structural engineers and builders *"What happens when fly ash is mixed into concrete?"* The question of course is meaningless. First of all we must have clear understanding which concrete properties we wish to enhance, and which we are prepared to sacrifice to a certain degree, when using fly ash. The end result will depend not only on the mineralogical and chemical composition of fly ash, but also on its grading, particle shape and impurities, and, most importantly, on the compatibility of physical, chemical and mineralogical properties of fly ash with the corresponding properties of Portland cement and sand used in the concrete mix [43], [44].

Exactly the same rule applies to the granulated blast furnace slag, rice husks, and other natural and artificial pozzolans. Silica fume, due to its very high surface area and fineness is particularly sensitive to the above factors.

Thus, there are no universal relationships providing optimum mix designs with Portland cement extenders. These mix designs must be derived for each specific application of concrete, and for each set of the available raw materials.

Impact on Greenhouse Gas Emissions

To produce one tonne of Portland cement clinker in a rotary kiln, some 100 kg. of fuel oil is used for the electric energy generation. To make Portland cement, clinker has to be ground to a fineness in excess of 300 m^2/ kg. Grinding requires between 35 and 40 kilowatt hours (i.e. 126 - 144 MJ) of electric energy per tonne of clinker, or about 10 kg of fuel oil. Apart from oil, pulverised coal and natural gas, or dual fuels, such as oil-coal, oil-natural gas, and coal-gas are commonly used for clinker making. An average CO_2 emission factor in the fossil fuel combustion process is of the order of 0.22 kg of carbon dioxide for every kilogram of fuel. The electrical energy requirement to produce one tonne of cement is of the order of 440 kWh or approximately 1,580 MJ [3]. Calcination of one tonne of $CaCO_3$ generates 440 kilograms of CO_2, according to the stoichiometry of this chemical reaction. Energy demand for the calcination of one tonne of clinker is about 490 kWh, or approximately 1,750 MJ.

Consequently, for every *tonne* of Portland cement produced, cement plants generate approximately *one tonne* of carbon dioxide - one of the main greenhouse gases!

This figure varies, depending on the type of kiln and the fuel used. For a *wet* process, common in the developing countries, it is near 1.1 tones of CO_2 per tonne of cement; for a more energy efficient *dry* process it is of the order of 0.9 tonnes of carbon dioxide per tonne of cement, but a *tonne per tonne* is a good average.

As a result, one cubic metre of good structural concrete, made with 320 kg/m^3 of Portland cement only, will generate some 0.41 tonnes of CO_2, emitted into the atmosphere. If 30% of Portland cement is replaced with fly ash, this figure is reduced to about 0.29 tonnes of CO_2, that is approximately 30% reduction of green house gas emissions, neglecting of course the carbon dioxide produced during the combustion of coal. With very reactive, highly pozzolanic and ultra-fine fly ash much higher Portland cement replacement (or replacement-addition) rates are possible, without reduction of the ultimate strength of concrete. In my experience, the reduction of CO_2 emissions to below 0.2 tonnes per cubic metre of concrete is possible, with only a slight effect on the rate of concrete strength gain, say if 90 in place of 28 day Characteristic Strength criteria is adopted [40].

Much higher Portland cement replacement rates are achievable with ground granulated blast furnace slag, and with *triple* or even *quadruple* blends, particularly if silica fume and High Range Water Reducing Admixtures are used.

Depending on the source, grinding of slag usually, but not always, requires slightly more than 40 kilowatt hours per tonne of ground material.

Even so, ground granulated slag manufacturing results in less that 11% of energy required to produce one tonne of Portland cement.

Optimising the composition of blended cements and curing methods, and, when possible allowing 90 day Characteristic Strength criteria in Structural Design, it should be possible to reduce the CO_2 emissions to less than 0.1 of a tonne per cubic metre of concrete.

HYDRAULIC CEMENTS MANUFACTURED USING INDUSTRIAL BY-PRODUCTS

Alternative Fuels

In addition, or even in place of the fossil fuels mentioned above [3], waste - derived fuels can be used in Portland cement manufacturing. Instead of natural gas, coke oven, pyrolysis and landfill gases are being used. Mineral, hydraulic, industrial oils, distillation residues and halogen-free spend solvents are also viable alternative liquid fuels. Pulverised coal can be supplemented and in some cases possibly replaced with tar, petroleum coke, sawdust, dried sewage sludge, some selected plastics, agricultural residue and car tyres. Steel-belted tyres in some cases may assist in the iron deficiency or a raw feed [3].

In 1995 the thermal energy consumption from alternative fuels in the European cement industry was about 10% [45]. In most cases replacement of fossil fuels with wastes reduces carbon dioxide emissions from clinker and cement manufacturing, the maximum reduction reported is of the order of 40%. However, in each particular case the energy efficiency of waste combustion in cement kilns and the quality, and particularly the consistency of manufactured cement should be closely monitored.

Waste introduced into the burning or calcining zone is combusted with a flame and gas temperatures approximately in the range of between 1,900° C and 2,100° C. This is sufficient to destroy all the hazardous organic components of waste. However, presence of trace elements, and particularly of heavy metals should be monitored [46], as well as the potential variations in the emissions to atmosphere [47].

Apart from dust, cement kilns emit into the atmosphere nitrogen oxide, sulphur dioxide, organic carbon, cadmium, thallium, beryllium, mercury and traces of arsenic, cobalt, nickel and lead.

Utilisation of Industrial Wastes in Manufacturing of Hydraulic Cements

Industrial wastes can also be used as a raw feed in manufacturing new types of hydraulic cements, which often results in the reduction of energy required for clinker formation, and thus in the lower emissions of CO_2 [3]. Since the late 1950-s theoretical foundations of new types of binders, produced by chemical or mechano-chemical activation of natural siliceous materials or industrial siliceous wastes were developed in several East European countries.

Chemical activation with caustic alkalis and with salts of alkaline metals produced so called *soilcement.* Sodium metasilicates or water glass were used to obtain new hydraulic cements. The hardening and the strength gaining processes in these binders can be enhanced and accelerated by a heat treatment, with the best results reported at the temperatures in the range of between 900°C and 1000°C [48]. As this process requires no calcination, and with the significantly lower processing temperature, as compared with the temperature demand during the Portland clinker formation, which is of the order of 1,400°C, it results in better than 50% reduction in the emissions of carbon dioxide, in contrast to even the most modern Portland cement plants.

By optimising the technology of manufacturing slag-alkaline cements, one day compressive strength of more than 50 MPa was achieved in a concrete based on this binder. One month compressive strength was of the order of 160 MPa. Currently, there is wide spread reported use of slag-alkaline cements in the countries, which were part of the old Eastern block [49], [50], [51], [52].

Another family of inorganic binders, closely related to slag-alkaline cements, which also challenge Portland cements, are geopolymers. These are mineral polymers which are based on inorganic silico-aluminates. One of several possible hardening mechanisms of alumino-silicate oxides involves chemical reaction with alkalis and alkali-polysilicates, yielding polymeric Si - O - Al bonds [53], [54]. An inorganic polymerisation reaction results in a three-dimensional structure, similar to that of zeolites.

Geopolymers are usually produced by blending three ingredients - calcined alumino-silicates, alkali-disilicates and granulated blast furnace slag or fly ash. Geopolymeric concretes were reported to reach compressive strengths of 20 MPa in 4 hours, and up to 100 MPa in one month [55]. Geopolymers were commercially introduced into Eastern Europe and the United States in the 1980-s, and are used for most construction purposes and also for the immobilisation of hazardous wastes [56].

As the geopolymers do not use $CaCO_3$ as a raw feed component, and also because chemical reaction of formation of alumino-silicates takes place at the temperatures of approximately 750°C, it is claimed that production of geopolymers results in between 0.1 tonne and 0.2 tonnes of CO_2, emitted for every tonne of this mineral binder [57].

Yet another alternative to Portland cement are the new types of binders, produced by mechano-chemical activation of seemingly inert siliceous materials, such as fly ash or silica sand. This process requires no calcination of clinker, and also no high temperatures for binder formation, and thus produces no carbon dioxide from either the combustion of fuel, or from the conversion of $CaCO_3$, essential in the Portland cement clinker manufacturing.

These cements are known as silica-water-suspension-binders. They are produced in specially designed and patented high intensity planetary mills. It was reported, that the best results were obtained after the concretes based on silica-water-suspension-binders were cured at temperatures of between 40°C and 120°C. Even allowing for these high curing temperatures, up to 90% reduction of carbon dioxide emissions, as compared with Portland cement production, can be achieved with these binders [3], [58].

WASTES AS REINFORCEMENT AND AS SPECIAL ADDITIVES IN CONCRETE

Waste as Continuous and Fibre Reinforcement and Additives for Self-Curing Concrete

Possibly the most common and one of the oldest in use of all the organic continuous reinforcing materials in Portland cement concrete is bamboo. It has large percentage of fibre, which is high in tensile strength, and it also has very good strength to weight ratio. Bamboo waste, as well as specially harvested plants have been used as concrete reinforcement in many developing, tropical and sub-tropical countries. Bamboo is generally used with low modulus concrete and requires special provisions for the end anchorage [59]. It is recommended, that bamboo main reinforcement in Portland cement concrete should be in the range of between 3.5% and 4.5%. Under the serviceability loads the reinforcement strains are then relatively small, and there is a large safety factor between the rupture and the serviceability loads [60]. The main weakness of bamboo, as indeed of any organic material, is its high water absorption potential, which ultimately leads to swelling and shrinkage. Yet another handicap of bamboo reinforcement is its relatively fast rate of decay in concrete [61]. To overcome the problems influencing long term durability, bamboo is often pre-treated, initially using sand-blasting to remove the outer skin, with subsequent soaking in the liquid sulphur. To prevent insect attacks, bamboo reinforcement can also be soaked in diesel oil [60], [61]. Field tests indicate, that over periods of up to 10 years in service, durability of bamboo reinforced concrete subjected to a mild environmental exposure is usually not a significant problem.

Many organic fibres, including wastes, are used as concrete reinforcement. These can be broadly classified as natural, which can be either of vegetable or animal origin, or synthetic [62]. Most natural or even synthetic fibres of organic origin have relatively low Modulus of Elasticity. The most common use for these fibres is to restrain plastic shrinkage cracking of fresh concrete. In hardened, structural Portland cement concrete the strains in these fibres are usually too big to effectively control drying shrinkage, thermal and creep induced cracking [63]. There are, of course, very strong synthetic fibres, which are manufactured specifically for the purpose of reinforcing brittle materials, such as concrete, but these are certainly not wastes. There are, still, some natural fibres of reasonably high Elastic Modulus, such as coir (4 to 6 GPa), jute (17 to 18 GPa), and sisal (35 to 60 GPa) [62], [64], [65]. However, in many developing countries, and particularly in the low-cost and temporary housing projects, bagasse fibre [66], coconut husk fibre [67], as well as many other natural organic fibres [62] have been successfully used for these special applications.

Some of the industrial wastes, such as relatively short steel wire off-cuts, provided they are of reasonably uniform size, can be used in structural Portland cement concrete, in place of the specially manufactured steel fibre reinforcement [68]. It is most important, that the off-cuts are free of impurities harmful to concrete. Thus, they have to be properly cleaned of oil and grease, usually with the caustic soda. In one of my unpublished R&D projects, greasy steel off-cuts actually caused reduction in the tensile strength and increase in the creep of concrete, reinforced with these fibres.

It is known, that special admixtures, which help to retain moisture inside the fresh Portland cement concrete, and then to slowly release it, assisting the long term process of cement hydration in hardened concrete, can improve its strength and durability.

Some waste materials, such as water saturated porous slag aggregate, can also assist with the self-curing properties of concrete. There are several student projects at the University of Wollongong, in which water saturated fibres from fly ash filter bags are used in concrete, to assess their effectiveness as the self-curing agents.

Wastes as Organic Additives and Binders

Some organic waste materials can be used as binder-extenders in bituminous concrete [69], or as binders in the paving materials in their own right [70]. Wood lignins were successfully used as extenders of bitumen, and lignins as well as waste oils as binders for the wearing courses in the low cost and temporary (military) pavements. Owen dry powder-like lignin is mixed into bitumen at about 135°C. At lignin additions of some 30% by weight of bitumen, the hardness, softening point and viscosity of the binder is increased. The blend has higher resistance to the fatigue failure, with the resistance to moisture and freeze-thaw damage virtually unchanged [69]. In yet another project a blend of epoxy resins, coal tar and phenol was used as a binder for special application concrete [71].

In Canada, among several other countries, there were periods of considerable over supply of sulphur. Although not strictly speaking a waste in accordance with the adopted definition, sulphur, during the low-price periods, requires some new methods of utilisation. This lead to the development and reasonably wide spread use of sulphur concrete [72], [73]. The sulphur concrete, very much like polymer impregnated concrete [74], is a specialised product. Its compressive and flexural strengths are of the order of 100 and 10 MPa respectively. The process technology is relatively simple and can be commercially viable, if the cost of sulphur is sufficiently low. Sulphur concretes are used when good chemical resistance, high impermeability, and good mechanical properties are required [73].

WASTES AND CONCRETE - POTENTIAL AND REAL ENVIRONMENTAL BENEFITS

Utilisation of Concrete for the Effective Waste Disposal

Concrete, under the broad definition of this term, can be an excellent material for the encapsulation of hazardous wastes. Fundamentally, two types of encapsulation are possible. One is a *condensed waste storage* for those wastes, which can be later re-processed and recycled, and the second is *dispersed waste disposal system*, in which low concentration of hazardous waste does not produce harmful effects in the encapsulating concrete, which therefore can be used as building and construction material [74]. The composition of Portland cement concrete can be optimised, using mineral and chemical admixtures, which in themselves can be waste products, for the encapsulation of each particular hazardous waste [1], [46], [74], [75].

However, the best results in most cases can be obtained with concretes based on hydraulic cements other than Portland, or on the organic cements and, sometimes with concretes based on the hybrid of two of the above types [1], [8], [74].

Slag-alkaline cements [50], [76], [77], geopolymers [55], [56], [78] and silica-water-suspension binders can be particularly effective for the long term immobilisation of hazardous wastes in concrete.

Reduction of Environmental Pollution and Abatement of Greenhouse Gas Emissions

Most of the existing waste management systems attempt to either eliminate processes which generate waste, or, as often is the case, if this is not possible, to find some commercial use for the waste of given physical and chemical characteristics and composition. There is of course a third alternative, which can often prove to be the most economically viable solution to the process of waste management. It is usually possible to modify the process which generates waste, without any detriment to its efficiency and possibly even enhancing this process, in such a way that the properties of waste will change, converting waste into a value added product. One example of this philosophy is the concept of *Total Fly Ash Management* [40]. It is theoretically and economically possible to select or modify the composition of pulverised coal in such a way, that the fly ash produced in power generating plants would have desirable characteristics for particular commercial applications. These properties would include an improvement in the pozzolanic activity of fly ash, intended for use in concrete, or provide a desirable carbon content for aggregate manufacturing, or a dramatic increase in the content of cenospheres in fly ash. Elimination of ash disposal dams and landfill sites, which are the potential cause for the contamination of land and water with chemical leachates, should improve the environment. Reduction in energy of manufacturing hydraulic cements and concretes, which are made from or contain fly ash, does assist the abatement of Greenhouse Gas emissions. *Total Fly Ash Management*, of course, is only one of many concepts, which can be just as economically viable.

The Greenhouse Gases which were included in the Kyoto agreement are - carbon dioxide, methane, nitrous oxide, hydrofluorocarbons, perfluorocarbons, and sulphur hexafluoride. Unlike the main gases that make up 99.97 per cent of the air - nitrogen, oxygen and argon, which are either diatomic or monatomic, Greenhouse Gases have three or more atoms. These gases can absorb infra-red radiation because their molecules do naturally vibrate at frequencies the same as the infra-red. It should be remembered that water vapour is also a Greenhouse Gas, which at present we have no means of controlling. The need to comply with the Kyoto Protocol would impose significant pressure on the cement and concrete industries world wide to reduce Greenhouse Gas emissions. Use of wastes as raw materials for the traditional and new cements and also as fuel-extenders should provide these industries with the sufficient means to meet the adopted targets.

CONCLUSIONS

Under the "business as usual" scenario, the International Energy Agency (IEA) predicts the world Primary Energy Demand to increase from approximately 8,000 million tonnes of oil equivalent (MTOE) in 1995 to over 13,000 MTOE by the year 2020 [79]. In 1997-1998 oil and natural gas accounted for just over 50% and coal supplied 22% of the energy used worldwide. About 17% of the world's electricity energy was produced by 441 nuclear plants, which of course do not generate Greenhouse Gas from the fuel combustion.

At the same time the United Nations Population Division assessed that the Earth's population would grow from the current estimate of nearly 6 billion people to 8 billion people in 2025, and to around 9.4 billion by the year 2050 [80]. This represents a considerable increase in the potential land, water and air pollution as well as the Greenhouse Gas emissions. Notwithstanding these trends, under the Kyoto agreement United States must reduce Greenhouse Gas emissions by 7%, and the European countries agreed to 8% reduction from 1990 levels by the year 2012. Japan agreed to a 6% cut, and Australia was given a special dispensation to actually increase the 1990 levels by 8%. Developing, and less developed countries were not included in the agreement. The impact of Germany's decision to close all Nuclear Plants, on the ability to comply with the Kyoto Protocol, is difficult to evaluate.

To achieve these objectives, it is essential to implement the concept of sustainable development. Sustainability is the ability to *maintain a desired condition over time.* It is recognised, however, that the concept of *a desired condition* can vary significantly, and that it is not only a function of ethnic, historical and economical characteristics for a common group of people, but also of the existing at the time environmental, political and possibly also cataclysmic, etc., factors. *Sustainable Development is a tool for achieving sustainability, and not a desired goal.* It should help humans to achieve and maintain healthy and productive life style, by integrating environmental management into all forms of the development processes. Utilisation of wastes in the cement and concrete industries is, of course, an essential part of such development [81].

REFERENCES

1. SAMARIN, A. Encapsulation of Solid Wastes from Industrial By-products, Keynote address, Intern. Conf. on Environmental Management, Geo-Water and Engineering Aspects, Wollongong, Australia, Eds. Chowdhury and Sivakumar, A A Balkema, Roterdam, 1993, pp 63-78.

2. HANSEN, T C. Recycling of Demolition Concrete and Masonry, RILEM Report No.6, (Technical Committee 37-DR6), E&FN Spon, 1992.

3. SAMARIN, A. Hydraulic Cements - New Types and Raw Materials and Radically New Manufacturing Methods, Leader Paper, Intern. Conf. Concrete in the Service of Mankind, Appropriate Concrete Technology, Dundee, Eds. Dhir and McCarthy, E&FN Spon, 1996, pp 265-279.

4. TERREL, R L, *et al.* Recycling of Existing Bituminous Materials for Rehabilitation of Pavements in Developing Countries, Intern. Conf. Materials of Construction in Developing Countries, Bangkok, Thailand, Eds. Pama, *et al,* 1978, Vol II, pp 567-581.

5. SCHUHBAUER, A. On the Properties and Application of Semi-Rigid Pavement Material, Residential Workshop, Materials and Methods for Low Cost Road, Rail, and Reclamation Works, Leura, Australia, Eds. Lee, *et al* W H Sellen, 1976, pp 405-437.

6. WILK, W, AND TSOHOS, G. Recycling of Concrete Pavements, Intern. Conf. Concrete 2000, Economic and Durable Construction Through Excellence, Dundee, Eds. Dhir and Jones, E&FN Spon, 1993, pp 1201-1209.

7. SAEKI, N, AND SHIMURA, K. Recycled Concrete Aggregate as a Road Base Material in Cold Regions, Intern. Conf. Concrete in the Service of Mankind, Concrete for Environmental Enhancement and Protection, Dundee, Eds. Dhir and Dyer, E&FN Spon, 1996, pp 151-156.

8. SAMARIN, A. Advanced Concrete Technologies - Opportunities for the Next Decade, Concrete Inst. of Aust. 15-th Biennial Conference, Concrete 91, Sydney, pp 103-111.

9. FRONDISTOU-YANNAS, S A. Recycled Aggregate as New Aggregate, Progress in Concrete Technology, CANMET, Ed. Malhotra, 1980, pp 639-684.

10. ELAZHARI, S A, *et al.* Recycling of Concrete at an Early Age, Intern. Conf. Concrete 2000, Economic and Durable Construction Through Excellence, Dundee, Eds. Dhir and Jones, E&FN Spon, 1993, pp 401-410.

11. SPEARE, P R S, AND BEN-OTHMAN, B. Recycling Concrete Coarse Aggregates and Their Influence on Durability, Intern. Conf. Concrete 2000, Economic and Durable Construction Through Excellence, Dundee, Eds. Dhir and Jones, E&FN Spon, 1993, pp 419-432.

12. HENDRIKS, Ch, F. Recycling and Reuse as a Basis for Sustainable Development in Construction Industry, Intern. Conf. Concrete on the Service of Mankind, Concrete for Environmental Enhancement and Protection, Dundee, Eds. Dhir and Dyer, E&FN Spon, 1996, pp 43-54.

13. COLLINS, R J. Increasing the Use of Recycled Aggregate in Construction, Intern. Conf. Concrete in the Service of Mankind, Concrete for Environmental Enhancement and Protection, Dundee, Eds. Dhir and Dyer, E&FN Spon, 1996, pp 73-80.

14. AJDUKIEWICZ, A B AND KLISZCZEWICZ, A T. Properties of Concrete with Rubble Aggregate from Demolition of RC/PC Structures, Intern. Conf. Concrete in the Service of Mankind, Concrete for Environmental Enhancement and Protection, Dundee, Eds. Dhir and Dyer, E&FN Spon, 1996, pp 115-120.

15. de VRIES, P. Concrete Re-Cycled: Crushed Concrete as Aggregate, Intern. Conf. Concrete in the Service of Mankind, Concrete for Environmental Enhancement and Protection, Dundee, Eds. Dhir and Dyer, E&FN Spon, 1996, pp 121-130.

16. di NIRO, G, *et al.* Recycled Aggregate Concrete (RAC): Properties of Aggregate and RC Beams Made from RAC, Intern. Conf. Concrete in the Service of Mankind, Concrete for Environmental Enhancement and Protection, Dundee, Eds. Dhir and Dyer, E&FN Spon, 1996, pp141-149.

17. UCHIKAWA, H AND HANEHARA, S. Recycled Concrete Waste, Intern. Conf. Concrete in the Service of Mankind, Concrete for Environment Enhancement and Protection, Dundee, Eds. Dhir and Dyer, 1996, pp 163-172.

18. OZKUL, M H. Properties of Slag Aggregate Concrete, Intern. Conf. Concrete in the Service of Mankind, Concrete for Environmental Enhancement and Protection, Dundee, Eds. Dhir and Dyer, 1996, pp 543-552.

19. MATHER, B. Mineral Aggregate for Concrete - Research Needed, Progress in Concrete Technology, CANMET, Ed. Malhotra, 1980, pp 57-110.

20. MONTGOMERY, D G AND NANG, G. Strength Properties of Concrete Containing Instant-Chilled Steel Slag Aggregate, Intern. Conf. Concrete for the Nineties, The Use of Fly Ash, Slag, Silica Fume and Other Siliceous Materials in Concrete, Leura, Australia, Eds. Butler and Hinczak, 1990, pp Mont. 1 to 14.

21. SUSTERSIC, J, *et al.* High Performance Concrete Using High Strength Artificial Aggregate, Intern. Conf. Concrete in the Service of Mankind, Concrete for Environment Enhancement and Protection, Dundee, Eds. Dhir and Dyer, E&FN Spon, 1996, pp569-578.

22. PINTO, T de P AND AGOPYAN, V. Recycling Construction Wastes as Raw Material for Low-Cost Construction in Brazil, Intern. Conf. Concrete in the Service of Mankind, Concrete for Environmental Enhancement and Protection, Dundee, Eds. Dhir and Dyer, E&FN Spon, 1996, pp 109-144.

23. De PAUW, C, *et al.* Reuse of Construction and Demolition Waste as Aggregate in Concrete, Technical and Environmental Aspects, Intern. Conf. Concrete in the Service of Mankind, Concrete for Environmental Enhancement and Protection, Dundee, Eds. Dhir and Dyer, E&FN Spon, 1996, pp 131-140.

24. TREVORROW, A. The Recycling of Construction Waste - a Planning Nightmare? Intern. Conf. Concrete in the Service of Mankind, Concrete for Environmental Enhancement and Protection, Dundee, Eds. Dhir and Dyer, E&FN Spon, 1996, pp 157-162.

25. BROWN, B V. Alternative and Marginal Aggregate Sources, Intern. Conf. Concrete in the Service of Mankind, Conctrete for Environmental Enhancement and Protection, Dundee, Eds. Dhir and Dyer, E&FN Spon, 1996, pp 471-484.

26. KIBRIYA, T AND SPEARE, P R S. The Use of Crushed Brick Coarse Aggregate in Concrete, Intern. Conf. Concrete in the Service of Mankind, Concrete for Environmental Enhancement and Protection, Dundee, Eds. Dhir and Dyer, E&FN Spon, 1996, pp 495-503.

27. MANSUR, M A. Crushed Bricks as Coarse Aggregate for Concrete, Intern. Conf. Concrete in the Service of Mankind, Concrete for Environmental Enhancement and Protection, Dundee, Eds. Dhir and Dyer, E&FN Spon, 1996, pp 505-514.

28. SAMARIN, A. Use of Fine Crushed Bottle Glass Sand and Partial Cement Replacement in Concrete, Intern. Conf. Materials of Construction for Developing Countries, Bangkok, Thailand, Eds. Pama *et al,* 1978, Vol I, pp 369-92.

29. RITCHIE, A G B AND TINGARI, M B. The Use of Waste Polystyrene Chips as a Lightweight Aggregate for Concrete, First Australian Conference on Engineering Materials, Uni. of New South Wales, Eds. Morgan and Welsh, 1974, pp 157-175.

30. SRI RAVINDRARAJAH, R AND TUCK, A J. Lightweight Concrete with Expanded Polystyrene Beads, Uni. Of Technology, Sydney, Civil Eng. Monograph No. C.E. 93/1 M.E., 1993.

31. AZIZ, M A, *et al.* Lightweight Concrete Using Cork Granules, Intern. Conf. Materials of Construction for Developing Countries, Bangkok, Thailand, Eds. Pama *et al,* 1978, Vol I, pp 181-189.

32. PARAMASIVAM, P AND LOKE, YO. Study of Sawdust Concrete, Intern. Conf. Materials of Construction for Developing Countries, Bangkok, Thailand, Eds. Pama *et al,* 1978, Vol I, pp 169-179.

33. ROSTAMI, H, *et al.* Use of Recycled Rubber Tyers in Concrete, Intern. Conf. Concrete 2000, Economic and Durable Construction Through Excellence, Dundee, Eds. Dhir and Jones, E&FN Spon, 1993, pp 391-399.

34. CONSUL, A M D AND MARTINEZ, HG. Alternative Building materials and Components for Concrete Construction in Cantiago de Cuba, Intern. Conf. Concrete in the Service of Mankind, Concrete for Environmental Enhancement and Protection, Dundee, Eds. Dhir and Dyer, E&FN Spon, 1996, pp 599-608.

35. HARRISON, AM, *et al.* The Use of Alternative Aggregates in Concrete, Intern. Conf. Concrete in the Service of Mankind, Concrete for Environmental Enhancement and Protection, Dundee, Eds. Dhir and Dyer, E&FN Spon, 1996, pp 543-552.

36. PIASECKI, J, *et al.* Concrete Waste Materials, Intern. Conf. Concrete 2000, Economic and Durable Construction Through Excellence, Dundee, Eds. Dhir and Jones, E&FN Spon, 1993, pp411-418.

37. LIAUTAUD, G. State of the Art Review of Practice and Experience on Low Cost Roads in Tropical Areas, Residential Workshop, Materials and Methods for Low Cost Road, Rail, and Reclamation Works, Leura, Australia, Eds. Lee *et al,* W.H. Sellen, 1976, pp 439-537.

38. BIJEN, J. Waste Materials and Alternative Products: Pro's and Con's, Intern. Conf. Concrete in the Service of Mankind, Concrete for Environment Enhancement and Protection, Dundee, Eds. Dhir and Dyer, E&FN Spon, 1996, pp 587-598.

39. MADEJ, J, *et al.* Investigation of Individual By-Products Considered for Use as Concrete Aggregate, Intern. Conf. Concrete in the Service of Mankind, Concrete for Environmental Enhancement and Protection, Dundee, Eds. Dhir and Dyer, E&FN Spon, 1996, pp 99-108.

40. SAMARIN, A. Total Fly Ash Management: from Concept to Commercial Reality, The Australian Coal Review, Issue 4, November, 1997 pp 34-37.

41. BIJEN, J M J M. Manufacturing Processes of Artificial Lightweight Aggregate from Fly Ash, The International Journal of Cement Composites and Lightweight Concrete, Vol 8, No.3, August 1986, pp 191-199

42. HARLEY, P. Neutralysis Technology Goes OS, Australian Environment Review, Vol 8, No.8, August 1993, p 3.

43. SAMARIN, A. Effect of Type, Compatibility and Proportions of Mix Constituents on the Durability of Reinforced Concrete, Japan-Australia Workshop on Concrete Durability, CSIRO, Highett, Victoria, Australia, November 28-30, 1988.

44. SAMARIN, A. Effect of Type, Compatibility and Proportions of Mix Constituents on the Corrosion of Reinforcement in Concrete, 2-nd International Seminar, Durability of Concrete - Aspects of Admixtures and Industrial By-Products, Gothenburg, Sweden, Chalmers Univ., Document D9, June 1989, pp 179-204.

45. CEMBUREAU, European Annual Review of Cement Industry and Market Data, No. 18, Brussels, Belgium, July, 1997.

46. SAMARIN, A. Encapsulation of Hazardous Heavy Metal Wastes in High Performance Concrete, Australian Academy of Technological Sciences and Engineering Journal Focus, No. 87, May - June 1995, pp 11-14.

47. AUSTRALIAN ACADEMY OF TECHNOLOGICAL SCIENCES AND ENGINEERING, Urban Air Pollution in Australia, Melbourne, Australia, 1997, Chapter 4, Large Stationary Sources, pp 157- 168.

48. GLUCHOWSKI, W D, *et al.* Slag-alkaline Cements and Concrete: Structures, properties, Technological and Economical Aspects of Use, Silicate Industries, No. 48, 1983, pp 197-200.

49. KRIVENKO, P V, *et al.* Slag-alkaline Polymer Cement Concretes, Intern. Conf. Concrete in the Service of Mankind, Appropriate Concrete Technology, Dundee, Eds. Dhir and McCarthy, E&FN Spon, 1996, pp 309-314.

50. RUNOVA, R F. Binders from Dispersion of Calcium Hydrosilicates, Intern. Conf., Concrete 2000, Economic and Durable Construction Through Excellence, Dundee, Eds. Dhir and Jones, E&FN Spon, 1993, Vol. I, pp 533-540

51. KIRILISHIN, V P. Constructive Chemically Resistant Concrete, Intern. Conf. Concrete 2000, Economic and Durable Construction Through Excellence, Dundee, Eds. Dhir and Jones, E&FN Spon, 1993, Vol II, pp 1013-1022.

52. SINGH, M AND VERMA, C L. Properties and Applications of New Cementitious Materials, Intern. Conf., Concrete 2000, Economic and Durable Construction Through Excellence, Dundee, Eds. Dhir and Jones, E&FN Spon, 1993, Vol II, pp 1313-1324.

53. DAVIDOVITS, J. Mineral Polymers and Methods of Making Them, U. S. Patent No. 434 9386, 1994.

54. DAVIDOVITS, J, *et al.* Geopolymeric Concretes for Environment Protection, Concrete International, Design and Construction, July 1990, Vol. 12, No. 7, pp 30-40.

55. DAVIDOVITS, J. Properties of Geopolymer Cements, Geopolymer Institute, Saint-Quentin, France, 1994.

56. van JAARSVELD, J G S AND van DEVENTER, J S J. Geopolymerisation: a Novel Technique for Treating Mining and Other Industrial Waste Materials, 2-nd Intern. Conf. On Environmental Management, Univ. of Wollongong, Australia, Eds. Sivakumar and Chowdhury, Elsevier, 1998, Vol II., pp 847-854.

57. DAVIDOVITS, J. Global Warming Impact on the Cement and Aggregates Industries, World Resources Review, Vol 6, No.2, 1993, pp 263-278.

58. MITIAKIN, P L AND ROSENTAL, O M. Refractory Materials Based on Silica-Water-Suspension Binders, Academy of Sciences of the USSR, Nauka Publishers, Novosibirsk, 1987, (in the Russian language).

59. DATYE, K R, *et al.* Engineering Application of Bamboo, Intern, Conf. Materials of Construction in Developing Countries, Bangkok, Thailand, Eds. Pama *et al,* 1978, Vol. I., pp 3-20.

60. GHAVAMI, K, *et al.* Structural Behaviour of Bamboo Reinforced Concrete Beams and Slabs, Intern. Conf. Concrete in the Service of Mankind, Appropriate Concrete Technology, Dundee, Eds. Dhir and McCarthy, E&FN Spon, 1996, pp 473-482.

61. FANG, H Y AND FAY, S M. Mechanism of Bamboo - Water - Cement Interaction, Intern. Conf. Materials of Construction in Developing Countries, Bangkok, Thailand, Eds. Pama *et al,* 1978, Vol. I., pp 37-48.

62. SARJA, A E. Tailoring the Properties of Concrete Structures with Appropriate Non-Ferrous Reinforcement, Intern. Conf. Concrete in the Service of Mankind, Appropriate Concrete Technology, Dundee, Eds. Dhir and McCarthy, E&FN Spon, 1996, pp 376-390.

63. SAMARIN, A. Concrete Shrinkage - Causes and Effects to be Considered in a Structural Design, Concrete in Australia, Conc. Inst. of Aust., Vol. 22, No. 1., April 1996, pp 18-23.

64. PAPAYANNI, I, *et al.* Low Cost Fibre Reinforced Cement Products by Using Inexpensive Additives, Intern. Conf. Concrete in the Service of Mankind, Appropriate Concrete Technology, Dundee, Eds. Dhir and McCarthy, E&FN Spon, 1996, pp 431-441.

65. JORILLO, P Jr AND SHIMIZU, G. Durability of Coir Fibre-Reinforced Concrete, Fourth CNMET/ACI Intern. Conf. on Durability of Concrete, Sydney, Australia, Rd. Malhotra, Supplementary Papers, 1997, pp 357-375.

66. RACINES, P G AND PAMA, P R. A Study of Bagasse Fibre-Cement Composite as Low Cost Construction Material, Intern. Conf. Materials of Construction for Developing Countries, Bangkok, Thailand, Eds. Pama *et al,* 1978, Vol. I., pp 191-206.

67. SINGH, S M. Coconut Husk - a Versatile Building Material, Intern. Conf. Materials of Construction for Developing Countries, Bangkok, Thailand, Eds. Pama *et al,* 1978, Vol. I., pp 207-219.

68. WALKUS, B R. Concrete Composites with Steel Micro-Reinforcement Subjected to Uniaxial Tension, Intern. Conf. Materials of Construction for Developing Countries, Bangkok, Thailand, Eds. Pama *et al,* 1978, Vol. I., pp 357-368.

69. TERREL, R L AND RIMSRITONG, S. Wood Lignins Used as Extenders for Asphalt in Bituminous Pavements, Intern. Conf. Materials of Construction for Developing Countries, Bangkok, Thailand, Eds. Pama *et al,* 1978, Vol. II., pp 583-596.

70. SAMARIN, A. Discussion on Paper by McDonald: Temporary and Engineering Roads and Pavements, Residential Workshop, Materials and Methods for Low Cost Road, Rail, and Reclamation Works, Leura, Australia, Eds. Lee *et al,* W.H. Shellen, 1976, pp 811-812.

71. KAWAKAMI, M, *et al.* Some Physical Properties of Resin Concrete as Construction Material, Intern. Conf. Materials of Construction for Developing Countries, Bangkok, Thailand, Eds. Pama *et al,* 1978, Vol. I., pp 275-283.

72. LOOV, R E, *et al.* The Use of Sulphur as a Binder for Soils and Waste Materials, Intern. Conf. Materials of Construction for Developing Countries, Bangkok, Thailand, Eds. Pama *et al,* 1978, Vol. I., pp 285-297.

73. MALHOTRA, V M. Sulphur Concrete and Sulphur Infiltrated Concrete - Properties, Applications and Limitations, Progress in Concrete Technology, CANMET, Ottawa, Canada, 1980, pp 583-637.

74. McLAREN, K G AND SAMARIN, A. Concrete - Polymer Composite Materials Produced by Radiation Polymerisation Techniques, First Aust. Conf. on Engineering Materials, Eds. Morgan and Welch, Univ. of New South Wales, Sydney, Australia, 1974, pp 137-155.

75. van EIJK, R J AND BROUWERS, H J H. Optimising the Portland Cement Matrix in Regard to Leaching Resistance, Second Intern. Conf. on Environmental Management, Univ. of Wollongong, Australia, Eds. Sivakumar and Chowdhury, Elsevier, 1998, Vol. II, pp 839-846.

76. DEJA, J AND MALOLEPSZY, J. Resistance of Alkali-Activated Slag Mortars to Chloride Solution, Third Intern. Conf. Fly Ash, Silica Fume, Slag, and Natural Pozzolans in Concrete, Trondheim, Norway, Ed. Malhotra, ACI SP-114, 1989, pp 1547-1553.

77. SAMARIN, A. Theory and Practice of Durable Concretes for Encapsulation of Hazardous Wastes, Invited Paper, Fourth CANMET/ACI Intern. Conf. on Durability of Concrete, Ed. Malhotra, Sydeny, Australia, Supplementary Papers, pp 833-855.

78. DAVIDOVITS, J. Geopolymers: Inorganic Polymeric New Materials, Journal of Thermal Analysis, No. 37, 1991, pp 1633-1656.

79. MACKEY, T. Energy: the Next Fifty Years, Australian Energy News, Issue 9, September 1998, pp 31.

80. ENCYCLOPAEDIA BRITANNICA, Yearbook of Science and Future, 1999, pp 308-311.

81. SAMARIN, A. The Anthropogenic Paradox - is Sustainable Future Possible? Keynote Paper, Second Intern. Conf. Environmental Management, University of Wollongong, Australia, Eds. Sivakumar and Chowdhury, Elsevier, 1998, pp 33-46.

CONCRETE BINDERS, MINERAL ADDITIONS AND CHEMICAL ADMIXTURES: STATE OF THE ART AND CHALLENGES FOR THE 21ST CENTURY

J S Damtoft

D Herfort

Aalborg Portland

E Yde

Nordisk Bygge Kemi A/S

Denmark

ABSTRACT. Although a wide range of concrete binders is available for specialist applications, by far the greatest volume is based on Portland cement. Mineral additions to Portland cement, particularly of artificial pozzolans such as fly ash, silica fume and blast furnace slag, and the use of more chemically specific admixtures continue to gain acceptance in many parts of the world. Although the hydrate mineralogy of the Portland cement systems is not radically altered by the presence of pozzolans, the content of C-S-H gel is generally increased at the expense of $Ca(OH)_2$ due to the reactive SiO_2 contents of the pozzolans, giving lower capillary porosities and therefore higher strengths and durability. Limestone fillers are also gaining acceptance and in some cases improve strengths and durability even at levels of replacement of up to 20%. For concretes where early strengths are less important than long term strengths and durability, mineral additions can be used to reduce CO_2 emissions and lower production costs. Systems containing two or more mineral additions combined with the optimum use of admixtures are creating opportunities for the development of high strength, highly durable concretes without compromising workability.

Keywords: Binders, Portland cement, Pozzolans, Admixtures, CO_2 emission.

Mr Jesper S Damtoft is Executive Manager, Research Development & Testing at Aalborg Portland, Denmark. His research interests include: Development of new cement and clinker types, recycling of waste and by-products as raw materials or fuels in the cement and concrete industry, novel cementitious materials and life cycle analyses.

Mr Elo Yde is Technical Manager of Nordisk Bygge Kemi A/S, Denmark, a member of SKW-MBT group. His responsibilities include research & development of admixtures for concrete and Quality Management.

Mr Duncan Herfort is Chief Geologist at Aalborg Portland. His responsibilities include all mineralogical aspects associated with raw material assessment, process and quality control, and concrete analysis.

INTRODUCTION

Concrete is the world's most important construction material. Modern construction in industry and of residential housing and infrastructure would be unthinkable without concrete. The volume produced is huge with approximately 5 km^3 produced annually [1]. Cement binders, additions and admixtures comprise a comparatively minor portion of modern concrete; typically 10-15%. Nevertheless, it is these minor constituents which are mainly responsible for the properties of fresh and hardened concrete, how much it costs to produce, and its impact on the environment. The main challenges faced by the concrete industry as it enters the 21st century include the need to

- reduce the impact on the environment
- improve durability
- compete with alternative building materials

against a background of increasing demand in the developing world, and little or no growth in the industrialised west.

The materials at the industry's disposal in order to meet these challenges are the binder, which is the topic of this paper, and the aggregate. The vast bulk of concrete binders consist of Portland cement with or without mineral additions and admixtures.

BINDERS

The term "binder" is normally used to define a material which, after mixing in a plastic state, generally with an aggregate, sets and hardens to form a solid mass. An hydraulic binder such as Portland cement is a finely ground substance which achieves a plastic and workable state on mixing with water and then reacts to form hydrated minerals. A wide range of binders exists for specialised applications such as Plaster of Paris for the gypsum product industry, Sorel cement for industrial flooring, phosphate cement and water glass for rapid set and acid resistance, and many more. In terms of volume, the great majority of binders used in the construction industry are calcareous which, for the purpose of this paper, are defined as cements whose compositions fall within the CaO-rich region of the C-S-A ternary phase diagram shown in Figure 1.

Lime and hydraulic lime cements (as well as lime cements strengthened by additions of Portland cement as for example specified by ENV 459-1 [2]) are today mainly restricted to masonry applications, although they had wider applications in the past. They were the forerunners of Portland cement leading up to the Industrial Revolution, with a wide range of applications including some large scale constructions such as the famous Eddystone Lighthouse [3]. Ancient pozzolanic cements had a similar compositional range to that given in Figure 1 for the hydraulic limes. High aluminate cements (HAC) are used for high temperature and chemical resistant applications, or where particularly rapid hardening is required [4]. The anhydrous and hydrate mineralogy of HAC is markedly different from the other calcareous cement types. The remainder of this paper will focus on Portland cement and its various derivatives.

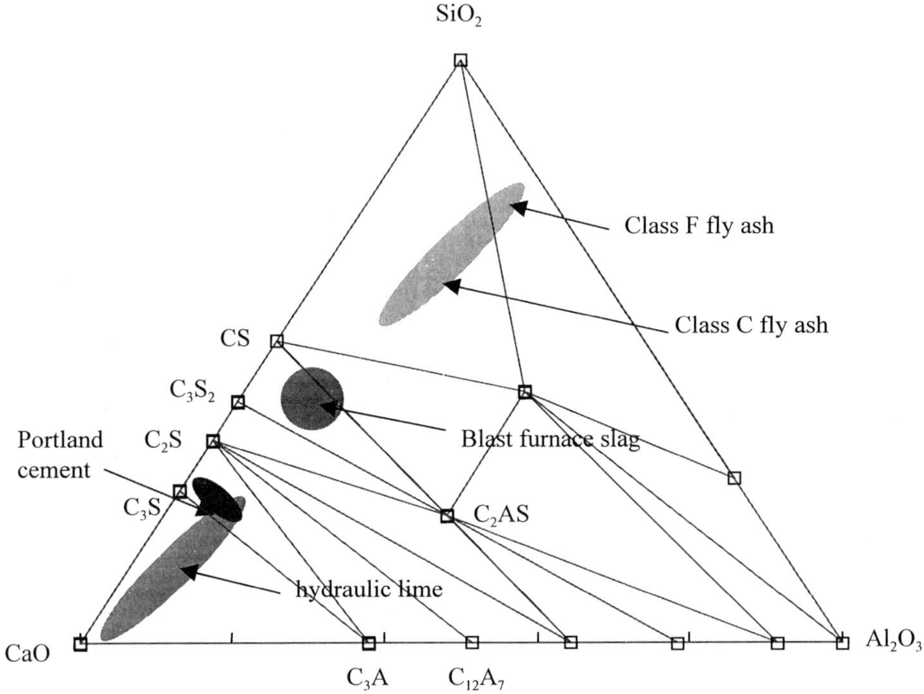

Figure 1 C-S-A phase diagram showing compositions of common calcareous cements

Portland Cement

In terms of chemical composition, modern Portland cement differs from the pozzolanic and hydraulic lime cements. The early Victorian Portland cements (termed by Blezard [3] as proto- and meso-Portland cements), had narrower and more consistent chemical compositions, and higher sulfate contents generally obtained by the controlled addition of gypsum for set regulation.

In terms of mineralogical composition, modern Portland cements have a much more well defined mineral assemblage, i.e. alite (C_3S), belite (C_2S), the aluminate phase (C_3A), the ferrite phase (C_4AF), in addition to approximately 1 % free lime. This is in contrast to the early Portland cements, hydraulic limes (which contained many non-equilibrium phases such as mayenite ($C_{12}A_7$) and gehlenite (C_2AS) due to inadequate blending of the raw feed and low burning temperatures), and lime-pozzolan cements in which the mineralogy is dependent on the raw materials (e.g. zeolites and volcanic glass in pozzolans of volcanic origin).

A feature common to all the cements plotted in Figure 1 is that they, at least in principle, form the same hydrate phase assemblage in the absence of carbonation, i.e. C-S-H, CH, AFt (usually ettringite) and AFm (usually monosulfate), an iron hydrate phase (probably an iron hydroxide gel containing Ca [5]) and the pore solution. The rate at which these hydrate phases are formed, and therefore the rate at which strength is gained through the reduction in capillary porosity, is dependent on the reactivity of the anhydrous minerals. The clinker phases in Portland cement are generally much more reactive than the minerals present in both natural and the artificial pozzolans described below. That is why mineral additions until now have generally not been included in rapid hardening cements.

The final hydrate assemblage is dependent on the combined chemical composition of all the hydraulic phases, and not on their mineralogy or reactivity. Portland pozzolan cements, which are generally richer in SiO_2 than pure Portland cements, result in a final hydrate assemblage with more C-S-H and less CH which, because of the comparative lower density of the former, results in lower capillary porosity at the same initial water/cement ratio. Greater durability due to lower permeability is also obtained in addition to higher ultimate strengths.

Apart from the capillary porosity which is determined by the chemical composition (as noted above) and the initial water/cement ratio, durability is also to a large extent dependent on the resistance of the individual hydrate phases to certain types of chemical attack. One of the best known examples of this is the inferior resistance of the aluminate-bearing phases, i.e. AFm and unreacted C_3A in the clinker, to sulfate attack. Because cements with a high content of Al_2O_3 can form more potentially expansive ettringite on reaction with sulphates from an external source, such as brine-rich groundwater, limits are placed on the C_3A content in cements specified as sulfate resisting (e.g. max. 5% C_3A for ASTM Type V and 8% for ASTM Type II) [6].

ADDITIONS

Blast Furnace Slag, Fly Ash, Silica Fume and Other Pozzolans

Mineral additions to Portland cement, added at the cement works or the concrete plant, have been common practice throughout most of this century with blast furnace slag in the heavily industrialised regions perhaps having the longest and most consistent track record. Over recent decades Portland cements containing fly ash from coal-fired power stations and silica fume from ferrosilicon production have gained increasing acceptance, whilst Portland cements containing natural pozzolans, rice husk ash and burnt oil shale are common in regions where these raw materials are readily available (Table 1). Replacement levels of Portland cement by blast furnace slag vary considerably with contents of well over 50 % by weight common in some regions. Fly ash typically replaces 10-30 % of the Portland cement although levels of 50-60 % have been advocated [7]. Where silica fume is added, it commonly comprises 5-10 % of the binder. ASTM standards exist for the use of natural pozzolans, fly ash, silica fume and blast furnace slag in concrete [8,9,10]. The only comparable European EN standard is for fly ash [11], although a standard for silica fume is currently in the course of preparation [12]. There are no plans for any EN standards for the use of blast furnace slag and natural pozzolans in concrete.

Table 1 Estimated availability of mineral additions worldwide [1, 13, 16, 17]

ADDITION	QUANTITY (MILLION. T/YEAR)
Fly ash	370
Granulated blast furnace slag	35
Silica fume	2
Rice husk ash	20
Portland cement (for comparison)	1,400

As shown in Table 1, fly ash is clearly the most abundant artificial pozzolan. Even so only approximately one third of it is used in the construction industry [13]. A large portion is produced in China and India, most of which could conceivably be used to advantage as cement and concrete production continues to increase in these countries [14]. The supply of fly ash in the industrialised countries, however, is expected to decline in the future, as coal-fired power stations are replaced by plants powered by renewable forms of energy [15].

Limestone Additions

Perhaps the most significant development in recent years has been the increased use of finely divided limestone, particularly in France, parts of Germany and in Denmark where advantage is taken of the European prestandard ENV 197 [18] allowing up to 35% $CaCO_3$. Many other countries permit the inclusion of up to 5% limestone as a minor addition.

ADMIXTURES

Admixtures are used to modify the physical properties of fresh concrete, the concrete's setting behaviour or the properties of the hardened concrete. The properties of the fresh concrete are modified by increasing its workability which can be used to reduce the water/cement ratio, reduce the cement content or increase the pumpability of the concrete. Other properties include the freezing temperature which can be lowered when using cold materials.

The setting and hardening of the concrete can be retarded or accelerated depending on the desired pot life, early strength development, etc. The properties of the hardened concrete can be modified by air entrainment which minimises the risk of freeze/thaw damage, by making it water-proof or more resistant to corrosion. Until recently, admixtures used in the concrete industry were usually bi-products of unrelated industries. Specific modifications for optimum use in the concrete industry were rarely possible.

Water-reducing admixtures were first introduced in 1935 as plasticizers based on ligno- and naphthalene sulfonates. In the early 1960's the first high range, or "super"-plasticizer, based on melamine and naphthalene formaldehyde, was introduced [19]. The dispersion mechanism of these superplasticizers in cement paste is best explained in terms of electrostatic repulsion between cement particles which are negatively charged by the adsorption of the polymer molecule onto the cement surface.

Table 2 Development of Admixtures in Concrete since 1885 [19]

1885	Patent of Calcium chloride
1925	Water-tightness admixtures
1930	Alumina powder
1932	Naphthalene-formaldehyde-con.
1938	Air entraining admixtures
1938	Accelerator and retarder
1939	Ligno-sulphonates
1950	De-foaming agent
1955	Freezing point lowering admixtures
1960	Melamine-formaldehyde condensate
1993	Polyacrylate
1997	Polycarboxlic ether

Over the past 10 years a new generation of polycarboxylate-based superplasticizers has been developed specifically for the concrete industry allowing far better control of key parameters such as strength, w/c ratio and slump retention. Structurally the new plasticizers are made up of a main flexible backbone, containing negatively charged carboxylic groups, and a large number of side chains (Figure 2).

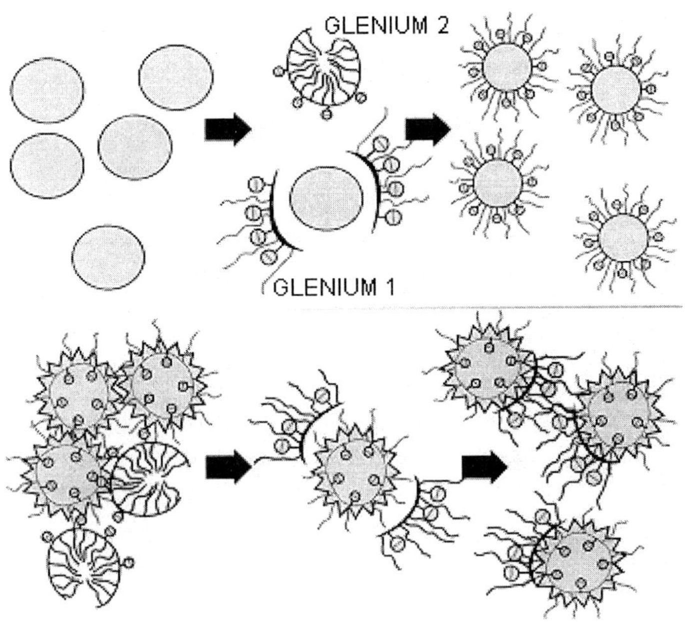

Figure 2 Mechanism of the action of polycarboxylic ether-based superplasticizisers, molecule 1 is activated immediately on mixing, whereas molecule 2, which requires a sufficiently high pH is usually activated after about one hour

Table 3 Water reduction from various water reducing admixtures [19]

Ligno-sulfonates	5-15 %
Melamine sulfonates	5-25 %
Naphtalene sulfonates	15-25%
Polyacrylics	20-30%
Polycarboxylic ether	25-40%

The dispersion mechanism involves two different types of repulsion forces between cement particles, i.e. electrostatic repulsion due to the presence of the negative charge of the carboxylic groups, and steric repulsion of the long side chains.

ADVANTAGES IN THE USE OF ADMIXTURES AND MINERAL ADDITIONS

Three main factors are currently driving trends in the use of admixtures and mineral additions in cement and concrete production:

1. Legislation and government incentives to reduce the impact on the environment, particularly CO_2 emissions.
2. Customer demand for more durable concrete.
3. Increased competition.

CO_2 Emissions

The most important political development over recent years has been the increasing commitment by industrialised countries to reduce CO_2 emissions. For the cement manufacturer, this is adding greater incentive to lower the energy input per unit clinker by improving kiln and grinding efficiency, to replace fossil fuel with CO_2-neutral waste materials such as used tires or plastic derived-fuels, and finally, to lower the clinker content of cement by replacement with pozzolans or limestone filler. In the final analysis, of course, it is not the amount of CO_2 emission per unit cement supplied by the cement manufacturer that counts, but the amount of CO_2 emission per unit concrete.

The CO_2 emission during the life cycle of four different types of concrete is shown in Figure 3. The contribution from the cement (which is calculated for ordinary Portland cement without mineral addition) is solely dependent on the cement content of the concrete, with the edge beam containing the highest content and the lightweight concrete containing the lowest. As noted in the previous section, the replacement level for mineral additions to Portland cement on a global basis is just over 10% compared to a potential replacement of approximately 30% based on present day availability. In other words, there is a potential for reducing the contribution made by the cement on CO_2 emission in concrete production by approximately 20%. However, whilst a weight for weight replacement of Portland cement by fly ash or slag may result in comparable or even improved long term strengths and durability, early strengths are reduced unless the overall content of the Portland cement is increased accordingly or the rate of clinker hydration enhanced by finer grinding.

For certain sections of the concrete industry such the precast and concrete product industries where high early strengths are required to minimised production costs, savings in Portland cement contents are probably minimal. For the ready mix producers, however, savings on Portland cement (and CO_2 emissions) are comparable to the level of mineral replacement and are therefore considerable.

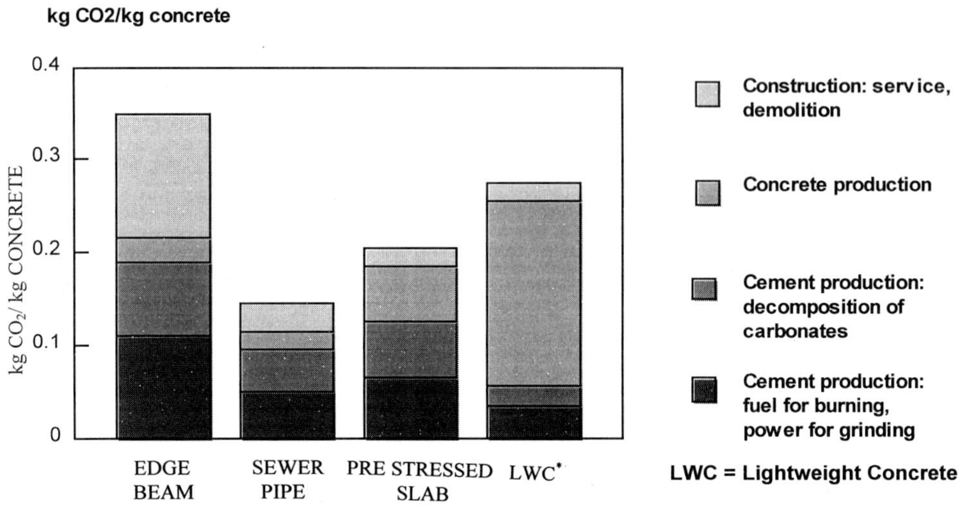

Figure 3 CO_2 emission during the life cycle of concrete [20]

Small additions of finely divided limestone of up to approximately 5% can result in comparable or improved strengths. Since the energy required to grind limestone is considerably less than that involved in the production of Portland cements, some reduction in CO_2 emissions is possible at these levels of replacement. At greater levels of limestone replacement, however, the limestone does not normally react and total cement contents in the concrete must be increased by an equivalent amount in order to maintain strengths, thus outweighing the environmental benefit. With special types of clinker, however, cements containing 20% limestone are possible without reducing strengths and porosity. This is described in more detail in the section of this paper on developments in Scandinavia and in a separate paper in these proceedings [21].

Durability

There is a general consensus nowadays that whilst prescriptive specifications on the composition of Portland cements that have been in place for many decades, such as the limit of 5% C_3A for ASTM type V cement, remain valid, the most effective method of improving durability is by reducing the w/c+m ratio (where m stands for mineral addition) [22,23]. The replacement of cement by artificial or natural pozzolans which are almost invariably richer in SiO_2 than Portland cement inevitably leads to lower capillary porosity at the same w/c+m ratio once most or all of the pozzolan has reacted.

Although this may take months or years, the end result is lower permeability and improved durability. It can be argued that the higher Al_2O_3 contents of fly ash and blast furnace slag compared to Portland cement may reduce resistance to sulfate attack [24].

The fact that limestone filler participates in the hydration reactions has been reported by a number of workers over recent years [25,26,27] despite other reports that it essentially acts as an inert filler [28]. It reacts mainly by forming monocarbonate ($C_3A.CaCO_3.12H_2O$), in addition to the hydrate phases already present. At replacements of approximately 5 wt% (depending on the aluminate content which is the main constraint on the amount of $CaCO_3$ that can react) capillary porosity is reduced, and from that point of view durability must be improved. Comparable or improved durability is confirmed by numerous performance tests [28,29,30]. Improved sulfate resistance from a mineralogical perspective has also been indicated [31], although concerns have been expressed that cool and moist conditions may lead to thaumasite formation [32]. As already noted in the previous section, except where special types of clinker are concerned, limestone replacement levels of more than 5-10% do not generally react and capillary porosities are therefore increased.

Water reducing agents such as plasticizers and superplasticizers are mainly used to reduce the water/cement ratio resulting in lower capillary porosity and therefore increased strength, lower permeability and improved durability. Improved workability may also indirectly improve durability by minimising the formation of placement related defects such as entrapped air. Air entraining agents generate air voids which, if sufficiently spaced, can relieve expansive pressures in the paste by filling up with the pore solution. They are primarily used to minimise the risk of freeze/thaw damage, but other expansive stresses such as those resulting from sulfate attack and alkali aggregate reaction may also be alleviated by air entrainment [33,34,35].

Increased Competition

The globalisation of the cement market, regional over capacity and the reduction in the cost of transporting bulk materials by sea have led to increased competition even in times of economic boom. This, and the increased harmonisation of cement standards has increased the perception of cement as an international commodity with nothing much to choose between Portland cements complying to the same specification apart from price. The replacement of Portland cement by pozzolans for applications where late strengths and durability are of primary importance, the use of limestone fillers, improved production processes, and the use of alternative fuels and raw materials in the production of cement increases competitiveness through a direct reduction in costs and improved environmental performance.

DEVELOPMENTS IN SCANDINAVIA

Reduction in Portland Cement Clinker Contents in Denmark

By replacements of fly ash and silica fume between 1977 and 1990

From the late 1970's until 1990, demand for higher concrete durability led to a general increase in average concrete strengths in Denmark.

This was achieved by replacing on average approximately 20% of the clinker with fly ash and silica fume, without increasing the binder content of the concrete [36]. Although the pozzolanic reaction of the fly ash and silica fume was largely responsible for this, improvements in other practices listed in Table 4 were also significant.

Table 4 Factors reducing cement content of Danish concrete 1977-90 [36]

Higher and more uniform cement strength	2-4%
Substitution of cement with fly ash	8-10%
Substitution of cement with silica fume	3-5%
Use of water reducing admixtures	1-3%
Improved process control at concrete plants	1-3%
Total estimated reduction in cement content	20%

By replacing with limestone filler between 1990 and 1999

Since 1990, further reductions in the clinker content in concrete have been achieved by the successful introduction of a Portland limestone cement based on mineralised clinker. The filler content of this cement is 14%. Since this cement enjoys a market share of approximately half of all the Portland cement sold in Denmark, the average reduction in the clinker content of concrete amounts to approximately 7%. That this has proved possible is due to the synergetic effect between clinker mineralised by CaF_2 and SO_3, and finely divided limestone filler, in which strengths are much higher than would normally be expected [21,25,37]. The new Portland limestone cement has significantly higher early as well as late strengths (EN 196 mortar 1 and 28 day strengths of 27 and 66 MPa respectively). This enables it to be used in applications, such as in the precast industry where high early strengths are a requirement, without needing to increase the cement content.

Since fly ash and silica fume additions are already common as indicated in Table 4, with the permitted levels of replacement shown in Table 5, it is quite common for the binder in concrete designed for all but the most aggressive exposure conditions to consist of four mineral constituents, i.e. limestone, fly ash, silica fume and as little as 50% Portland cement.

100 year guaranteed service life

An example of the way in which mineral additions and admixtures have been used to produce highly durable concrete is the recently completed Great Belt bridge and tunnel connection between the Danish islands of Fyn and Sjælland. Among other things, the contractor was required to guarantee a 100 year service life. The mix design for the bridge is shown in Table 6.

Further specifications include a slump of 180 mm (std. < 30 mm), a slump loss of < 50 mm after 1 1/2 hours, an air void content of 6.5% (std < 1.5%) with a spacing factor of < 0.20 mm, and a specified limits for autogenous and thermal cracks.

Considerable effort was employed in finding a compatible combination of admixtures in order to meet these requirements. Among other things, a water reducer/retarder was needed that did not adversely affect air entrainment. Once the correct combination was developed it proved possible to maintain sufficient workability for the specified period, although in some cases it was necessary to re-dose superplasticizer during use.

Table 5 Fly ash and silica fume according to the Danish concrete pre-standard [38]

EXPOSURE CLASS	(FA+SF)/(FA+SF+C)	SF/(FA+SF+C)
Passive	no requirements	no requirements
Moderate	maximum 35%	maximum 10%
Aggressive	maximum 25%	maximum 10%
Extra aggressive	maximum 25%	maximum 10%

FA: Fly ash, MS: Micro silica, C: Cement

Exposure classes

Passive:	Dry or subterranean exposure with no risk of damage due to freeze/thaw, chemical attack or corrosion of reinforcement
Moderate:	Outdoor exposure, but with no risk of freeze/thaw combined with water saturation, and with no significant external source of alkalis, sulfates, chlorides or other potentially deleterious components.
Aggressive:	Outdoor exposure, with significant risk of freeze/thaw combined with water saturation, or a significant risk of attack by of alkalis, sulfates, chlorides or other potentially deleterious components.
Severe:	Option available in which additional control measures may be required where exposure is judged to particularly aggressive.

Table 6 Mix design for the Great Belt Bridge project (kg/m^3) [39]

Cement (low heat, sulfate resisting, low alkali)	325
Fly ash (class F)	40
Silica fume	40
Air entraining agent	1.0
Plasticizer/retarder	1.7
Superplasticizer (melamine)	7.3
Maximum water content	130
Maximum water/cement ratio	0.33

The low heat, low alkali, sulfate resisting cement was chosen to minimise the risk of thermal and autogenous cracking, alkali aggregate reaction and sulfate attack, respectively. Apart from improving workability, the silica fume and fly ash were designed to increase ultimate strength and durability, mainly by reducing the ultimate capillary porosity, although the Al_2O_3 content of the fly ash is also expected to reduce chloride diffusion.

The technology applied in the Great Belt project was further developed for the Sound Link Project - a major tunnel-bridge construction linking Denmark and Sweden, which is currently under construction. Some of the specifications for the Sound Link concrete are listed in Table 7. In order to obtain a concrete with a relatively high degree of workability and stability, the Sound Link Tunnel Contractors I/S chose a superplasticizer based on melamine, in combination with a stabiliser (Melstab 21) and an additional retarder. Initial slump was typically 220 - 250 mm, being reduced to approximately 170 - 200 mm after pumping. Typical mix designs for the Sound Link concretes are shown in table 8.

Table 7 Specifications for the Sound Link concrete, Types A and B [39]

CONCRETE DEMANDS	TYPE A: TUNNEL SEGMENTS	TYPE B: RAMPS
General condition	- High degree of workability - Adjustable setting time - Precise density - Quick strength development	- Normal workability - Frost resistant
Water/cement ratio	< 0.40	<0.45
Cement content	> 275 kg/m^3	
Fly ash content	< 15 % of total powder	
Silica fume content	< 5 % of total powder	
Total powder	< 340 kg/m^3	
Compressive strength	40 MPa	35 MPa
Air content	no demands	> 3.0 vol. %
Frost resistance SS137244	no demands	< 1.0 kg/m^2

Table 8 Mix design for Sound Project Concrete types A and B (kg/m^3) [40]

	TYPE A	TYPE B
Cement	324	295
Fly ash	52	50
Silica fume	24	30
Air entrainment	-	1.0
Retarder	0.4	-
Plasticizer	-	0.9
Superplasticizer	10.5	5.4

The superplasticizers used for for types A and B are Melstab 21 and Melment L1504x respectively.

Self-compacting Concrete

Self-compacting concrete is a relatively new type of high performance concrete with very high workability necessary for placement in complicated structures such as tightly spaced reinforcement. This is made possible by a new generation of superplasticizers allowing much lower water/cement ratios than was previously possible. This technology has already been successfully applied in several European constructions.

As an example, three bridges were constructed in Sweden in 1998 using self-compacting concrete with a slump of 600 - 700 mm, and a 28 day compressive strength of 70 MPa. Part of the mix design is shown in table 9. Performance testing showed the concretes to be freeze/thaw resistant. Visual analysis including thin section microscopy revealed a dense homogeneous structure with none of the flaws characteristic of poor compacting and placement

Table 9 Part of mix design for Swedish bridge unit using self-compacting concrete [41]

MIX CONSTITUENT	kg/m^3
Cement	415
Limestone Filler, max 0.5mm	180
Superplasticizer, Glenium 51	3.3
Air Entrainer	0.04

w/c ratio = 0.40

Interest in self-compacting concrete is not restricted to major constructions. There is also considerable interest in the precast industry, where self-compacting concrete can reduce or even completely eliminate the need to vibrate. This has obvious benefits in reducing costs and improving workmanship as well as working conditions. Table 10 shows some typical mix designs for self-compacting, precast concrete.

Table 10 Mix design for self-compacting concrete used in pre-cast concrete [42]

ADMIXTURES (% OF POWDER)	CEMENT (kg/m^3)	FLY ASH (kg/m^3)	W/C RATIO	SLUMP FLOW	COMPRESSIVE STRENGTH (MPa)	
					1 Day	28 Days
0.6	242	72	0.50	500	-	53
0.9	330	80	0.39	600	19	55
1.0	320	60	0.41	670	21	62

The admixtures used in these mix designs are all second generation superplasticizers

The advantages of using mineral additions such as fly ash and silica fume described in earlier sections also apply to self-compacting concrete further improvements in performance.

CONCLUDING REMARKS

Concrete is a cost-effective and relatively environmentally friendly construction material. However, it does not exist in a vacuum and must remain competitive if it is to maintain its dominance as the preferred construction material into the 21st century.
The challenge faced by the concrete industry is to reduce the environmental impact of concrete throughout its life cycle, from the quarry to the end of the concrete's service life, whilst at the same time improving durability and cost efficiency.

The concrete binder based on Portland cement, mineral additions and chemical admixtures will remain the most effective way of meeting this challenge for the foreseeable future. The finite availability of artificial pozzolans, such as fly ash and blast furnace slag, however, is likely to pave the way for alternative replacement materials such as limestone filler.

REFERENCES

1. JAHREN, P. Har bransjen tatt problemet på alvor. BETON Gindustrien, No. 4, 1998, pp 26-30. (In norwegian)
2. ENV 459-1:1994:E, Building lime - Part 1: Definitions, specifications and conformity criteria, 1994.
3. BLEZARD, R G. "The history of Calcareous cements" in Lea's Chemistry of Cement and Concrete, 4 th edition Ed. P.C.Hewlett, Pub. Arnold, 1998, pp 1-23.
4. TAYLOR, H F W. Cement Chemistry, Pub. Thomas Telford, 1997, pp. 295-313.
5. EMANUELSON, A, AND HANSEN, S. Distribution of Iron Among Ferrite Hydrates. Cem. and Conc. Res, Vol. 27, No. 1, 1997, pp 1167-1177.
6. ASTM C 150-97, Standard Specification for Portland Cement.
7. BILODEAU, A, and MALHOTRA, V M. High-Volume Fly Ash System: The Concrete Solution for Sustainable Development. In Compilation of Papers, Three-Day CANMET/ACI Symposium on Sustainable Development of the Cement and Concrete Industry, October 21-23 1998, Ottawa, pp. 193-214
8. ASTM C 612-97, Standard Specification for Coal Fly Ash and Raw or Calcined Natural Pozzolan for Use as a Mineral Admixture in Concrete, 1997.
9. ASTM C 989-95, Standard Specification for Blast-Furnace Slag for Use in Concrete and Mortars, 1995.
10. ASTM C 1240-97, Standard Specification for Silica Fume for Use as a Mineral Admixture in Hydraulic-Cement Concrete, Mortar and Grout, 1997.
11. EN 450:1994E, Fly Ash for Concrete - Definitions, Requirements and Quality Control, 1994.
12. pre EN 13263:1998, Silica Fume for Concrete - Definitions, Requirements and Quality Control, 1998.
13. MANZ, O E. Worldwide Production of Coal Ash and Utilisation in Concrete and Other Products. Fuel, Vol. 76, No. 8, pp. 691-696, 1997.

14. MEHTA, P K. Concrete technology for sustainable development - An overview of essential principles. In Compilation of Papers, Three-Day CANMET/ACI Symposium on Sustainable Development of the Cement and Concrete Industry, October 21-23 1998, Ottawa, pp. 1-14.

15. DAMTOFT, J S. Use of Fly Ash and Other Waste Materials as Raw Feed and Energy Source in the Danish Cement Industry. In Compilation of Papers, Three-Day CANMET/ACI Symposium on Sustainable Development of the Cement and Concrete Industry, October 21-23 1998, Ottawa, pp. 95-105.

16. World Cement Annual Review 1997. World Cement, Vol. 28, No. 7, July 1997, p. 4.

17. MORANVILLE-REGOURD, M. Cements Made From Blastfurnace Slag. In HEWLETT, P C (ed.) Lea's Chemistry of Cement and Concrete. Fourth Edition. Arnold, London, pp. 633-674.

18. ENV 197-1, Cement - Composition, specifications and conformity criteria - Part 1: Common cements, October 1992.

19. KJÆR, U. Admixtures for Concrete, a Handbook. BKI Report 1, 1982.

20. Industry Analysis, Concrete - Cleaner Technology in Concrete Production. Environmental report No. 350, Danish Environmental Protection Agency, 1997. (In danish).

21. RASMUSSEN, S, AND HERFORT, D. Mineralised Cement, Separate peper in these proceedings, Conference 1: Modern Concrete Materials: Binders, Additions and Admixtures, Basic Portland Cement.

22. SKALNY, J, AND PIERCE, J S. Sulfate Attack Issues, Proceedings of the "Quebec Seminar on Sulfate", October 5 & 6, 1998, to be published as "Materials Science of Concrete, Special Volume: Sulfate Attack Mechanisms (J. Marchand & J. Skalny, Eds.) by the American Ceramic Society.

23. Guide to Durable Concrete (1992), section 2.2, ACI 201.2R-92.

24. LOCHER, F W. The problem of Sulfate resistance of slag cements (in German), Zement-Kalk-Gips, Vol. 55, No. 9, 1966, pp. 395-401.

25. KLEMM, W A, AND ADAMS, L D. An Investigation of the Formation of Carboaluminates. in Carbonate Additions to Cement, Klieger/Hooton, Eds. ASTM STP 1064, 1990, pp. 60-72.

26. BORGHOLM, H E, HERFORT, D, AND RASMUSSEN, S. A new blended cement based on mineralised clinker, World Cement vol. 8, pp. 27-33, (1995).

27. INGRAM, K D, AND DAUGHERTY, K E. A review of Limestone Additions to Portland Cement and Concrete. Cement and Concrete Composites, Vol. 13, No. 3, 1991, pp. 165-170.

28. SIEBEL, E, AND SPRUNG, S. Influence of Limestone in Portland Limestone Cement on the Durability of Concrete (in German). Beton, Vol. 41, No. 3, 1991, pp. 113-117.

29. RANC, R, MORANVILLE-REGOURD, M, COCHET, G, AND CHAUDOUARD, G. Durability of Cements with fillers, 2nd- CANMET/ACI International Conference on Durability of Concrete, Eds V.M.Malhotra, 1991, Vol. II, pp. 1239-1257.

30. SCHMIDT, B. Cement with interground additives - Capabilities and Environmental Relief, Part 2, Zement-Kalk-Gips, Vol. 45, No. 6, June 1992, pp. 296-301.

31. SOROKA, I, AND SETTER, N. Effect of Mineral Fillers on Sulfate Resistance of Portland Cement Mortars. Durability of Building materials and Components, ASTM STP 691, P.J. Sereda and G.G. Litvan, Eds., American Ceramic Society for Testing and Materials, 1980, pp. 326-335.

32. BENSTED, J. Scientific Background to Thaumasite Formation in Concrete. World Cement, November, 1998, pp. 102-105.

33. YANG, Q, WU, X, AND HUANG, S. concrete deterioration due to physical attack by salt crystallisation. Proc. 10th ICCC, Gothenburg, 1997, 4iv032 5p.

34. STARK, J, AND BOLLMANN, K. Ettringite Formation - A durability Problem of Concrete Pavements, Proc. 10th ICCC, Gothenburg, 1997, 4iv062 8p.

35. JENSEN, A D, CHATTERJI, S, CHRISTIANSEN, P, AND THAULOW, N. Studies of Alkali-Silica Reaction – Part II; Effect of Air Entrainment on Expansion, Cem. and Conc. Res, Vol. 14, No. 3, 1984, pp 311-314.

36. Cementindustrien og CO_2-afgiften. Aalborg Portland, 1990, 20 p.

37. BORGHOLM H E, AND DAMTOFT, J S. European Patent, EP 0640062B1, 1996.

38. DS 481:1998, Danish pre-standard for composition of concrete and concrete materials, 1998.

39. BAEKMARK, HANSEN AND REICHERT, "Stabilized Air Void System in Superplasticized Concrete", ACI SP-144, Concrete Technology Past, Present, and Future, pp.177-190, 1994.

40. LUNDBERG L. "Øresundstunnel - Performance and Demands on Concrete" (in Danish), Dansk Betondag 1997, Concrete Society of Denmark.

41. PETERSSON, Ö. Applications of Self-Compacting Concrete for Bridge Castings. Swedish Cement and Concrete Research Institute, 1998.

42. Self Compacting Concrete with Glenion 51, NBK unpublished internal report (in Danish), 1998.

READY MIX CONCRETE: AN OLD-NEW MATERIAL FOR THE 21ST CENTURY

F Biasioli

Polytechnic of Turin

Italy

ABSTRACT. Ready Mixed Concrete, a protagonist of the construction industry, is almost one hundred years old. Moving on from what has happened during the eighties and nineties, the paper outlines the foreseeable future of Ready Mixed Concrete from different points of view, covering mix design, production methods, equipment, customer-producer relationships, environmental impact and the European standards and regulations.

Keywords: Concrete, Ready mixed concrete, EN206, CPD, Recycled concrete, Factory production control, High performance concrete, GPS, Admixtures, Additions.

Professor Francesco Biasioli a civil engineer, is a Research Professor of Building Materials at the Department of Structural Engineering, Polytechnic of Turin, Italy. He specialises in concrete and reinforced concrete related matters. Professor Biasioli serves on many Technical Committees and as from 1997 is at present the Secretary General of ERMCO, the European Ready Mixed Concrete Organisation.

INTRODUCTION

Since the beginning, the objective of the ready mixed concrete industry was to supply the construction industry with quality fresh concrete. The ready mixed concrete industry's capacity for targeting the users needs with a competitive product, having the specified properties and suitable for the intended use, is one of the reasons why the use of ready mixed concrete has increased in every country throughout recent years.

The paper describes how the ready mixed concrete industry was born and has developed, and tries to identify the current situation and the foreseeable new challenges. An ideal profile of the industry at the beginning of the new century is also traced.

THE HISTORY OF READY MIXED CONCRETE

The history of concrete, "*...species of sand mixed with lime and rubble..*(which)*.. hardens as well as under water as in ordinary buildings..*" [1] dates back 10,000 years. The oldest "man-made stone" known today appears to be in the large area of compacted ground consisting of sandy aggregate, limy binder and water near Yiftah'el, in Israel's Lower Galilea. Babylonians and Assyrians used a mixture of clay, lime and water, Phoenicians and Egyptians mixed ground bricks and lime. Roman engineers built various magnificent buildings and civil structures, many of which still exist today.

Even if concrete is a thousand years old, it is only in the last quarter of the 19th century that the first idea of "ready-mixed" concrete became known. In 1872 a British engineer, Mr. Deacons, is quoted as having said "*undoubtedly, it should be regarded as a big advantage to deliver concrete or mortar ready to fit into the construction area*". Even if he was the first to envisage the efficiency of operating plants in order to supply concrete and mortar in the quantity and with the characteristics needed, the Germans claim, with some reason, that they were the first to actually turn this idea into reality [2].

Since 1875 damp sand and hydrated lime, a type of lime-sand mortar, was carried by horse and cart to building sites in Berlin. Apart from Deacon's words, this was probably the incentive which pushed the Hamburg based Architect Jurgen Heinrich Magens to start an intensive search for suitable ways of producing off-site and transporting fresh workable concrete over fairly long distances.

According to the records of what was then known as "the German Committee for Reinforced Concrete", Magens, born in 1857, was the responsible German "expert" in matters such as concrete and building materials. After having established his company in 1896, and knowing from experience that concrete does not harden at frosty temperatures, he first patented the idea of cooling concrete. January 10th, 1903, was the date of his first "Imperial Patent", and this can be regarded as the birthday of ready mixed concrete.

Magens realised very soon that, instead of cooling fresh concrete to a freezing point, the same result could be obtained by simply storing concrete materials in a cool place and/or by sprinkling aggregates with cold water - another Patent was then obtained. Some years later, in 1907, he finally discovered that the available time for transportation could be prolonged not only cooling the fresh concrete, but also by vibrating it during transport.

As he demonstrated that vibration also improved the characteristics of concrete, he obtained a further new Patent. Ready mixed concrete was definitely identified and some plants were set up in Hamburg and Berlin.

The first delivery of ready mixed concrete outside of Germany was effectively done in the United States just before the First World War. The increasing availability of special transport vehicles, supplied by the new and fast growing automobile industry, played a positive role in the development of the ready mixed concrete industry. In Europe, ready mixed concrete was produced in 1926 in Denmark, which is also the year when the first transit mixer appeared in the United States, and since the early thirties in the United Kingdom [3].

After the Second World War the ready mixed concrete industry spread all over Europe and now, with more than 13,000 plants and 350 million of m^3 of production, it supplies the largest part of fresh concrete used in the European civil engineering works. But what are the challenges that the industry has to face at the beginning of a new century?

FROM NATIONAL REGULATIONS TO EUROPEAN STANDARDS

Buildings and civil works have to perform safely and protect both the products and lives of users. In all countries, building legislation deals with construction design, workmanship and products. The Common European market means that the building sector, as other sectors, has to develop the transfer of many standardisation activities from a national to a supra-national, European level.

The importance of standardisation for increasing competitiveness in the European industry has been expressed in official documents by the European Commission, in particular in the so-called "New Approach" Directives. The Directives are based on the idea that, when specifying product characteristics, compulsory legislation should be kept to a minimum so as to favour a broader use of Standards as agreed by the National Industry Associations. This would result in an increase in both the Quality of Product and the competitiveness of the industry.

The Construction Products Directive 89/106 (CPD) issued in 1988 [4] is based on this approach. This Directive is aimed at establishing a network of harmonised European Standards for construction products belonging to 33 major families.

As a major construction material, concrete is deeply involved in the standardisation process, together with its constituents - cement, aggregates, admixtures, additions. The final version of EN206, the new European Standard for concrete, will be issued in late 1999 after almost twenty years of preparatory work [5]. This new version of the standard, a major step forward when compared to its preliminary release ENV206, will be applied to all types of concrete, both ready and site mixed, or used for production of precast structural components for buildings and civil engineering structures.

The standard incorporates rules for the use of constituent materials that are covered by other European standards, but concrete is a construction product that is largely based on materials that are available locally. Based on local experience, by-products of industrial processes, recycled materials etc. are in current use in several countries.

Until European specifications for these materials are available, EN206 does not provide rules for their use, instead it refers to national standards or provisions valid i the place of use of the concrete.

The standard has to be applied in Europe under different climatic and geographical conditions, which is why improved definitions of the so called "exposure classes", i.e. classes related to environmental conditions which can cover the different national situations, have been implemented into the standard.

Durability requirements for concrete composition, higher strength classes, improved consideration of additions, improved conformity criteria and a clear outline of what is required for factory production control are but some items which characterise the new standard. In order to avoid confusion and assign responsibility for any relevant decision, the standard clearly identifies tasks for the concrete specifier, producer and user, and the necessary exchange of information between the different parties.

Even if concrete will be identified, prescribed and controlled in a uniform way throughout Europe, it is impossible, at the moment, to foresee the real effect of the introduction of EN206 on the European ready mixed concrete industry. It has to be taken into consideration that, at present, the ready-mixed concrete industry is made up largely by small and medium enterprises, with only a few large groups extending their activities in several countries. Very small companies may have difficulties in following up the developments of legislation and technical regulations if these differ greatly from what they have used for in the past.

FROM TECHNICAL REGULATIONS TO QUALITY SYSTEMS

The development of updated cement and concrete standards has been set up in parallel with the introduction of quality systems within the concrete industry. The implementation of a quality system can improve the industries competitiveness enhancing production, reducing delays and waste, and contributing generally to a more motivating and successful working environment. Starting with the USA, quality systems were taken up by the Japanese during the 60s, and are now used in several industrial sectors.

The ready mixed concrete industry is far different from other building industries, but some "quality based" concepts are generally accepted:

1. at each level of the production process responsibilities have to be set out, and each activity has to be defined using appropriate procedures which clearly identify tasks and responsibilities;

2. because a number of different materials can be used in order to produce a large number of different mixes in a single plant, Factory Production Control is of major importance.

As defined in EN206 [6], production control comprises "...all measures necessary to maintain the quality of the concrete in conformity with specified requirements...". In order to be effective, production control has to take into account the kind and size of production and equipment, together with the procedures and rules in use at the place of use of concrete.

The practice of FPC certification is increasing in all countries and it could become a mandatory requirement for concrete attestation of conformity in a the future harmonised revision of EN206.

In many European countries producers already operate production control systems which, in all probability, will result in concrete conforming to the implicit and expressed requirements [3]. For these producers a further move can be envisaged, towards a stricter adherence to the requirements of the EN ISO 9000 series of standards.

READY MIXED CONCRETE: THE GREEN PRODUCT

Erecting civil works has an impact on the natural environment. Building materials are produced transforming natural resources and the production process itself causes emissions in air and water. This area is, sometimes indirectly, affected by many EU legislative initiatives. Laws and regulations on air and water quality, on waste management and in general landscaping issues, reflect the general concern about the quality of our living environment and tend to re-direct the industrialisation process.

Since 1960 the European Union has adopted over two hundred environmental Directives. The driving force is that any possible waste has to be avoided, but, if this cannot be realised, the waste has to be re-used, and if, and only if, the reuse is impossible, it has to be disposed.

The Water Policy Framework Directive, the Landfill of Waste Directive [7] and especially the 1993 Fifth Environmental Action program have set long term objectives for the Community policy, trying to shift the focus of environmental policy in relation to the industry from a prescriptive to a co-operative approach. Industry is now considered not only as a part of the environmental problem, but also as a part of the solution.

It is not generally recognised that the industrial production of fresh concrete has a low impact on the environment. However, both the ready-mixed concrete and the quarrying industries in all countries have been forced to improve the way they operate, i.e. recovery, re-use of waste water and other industrial by-products, reduction of noise and dust levels, re-use of demolition materials etc.

A production process that respects the environment can create a competitive advantage by using resources more effectively. In the case of the ready mixed concrete industry:

- processing operational waste water and its internal reutilization are no longer an objective but a reality for many plants all over Europe;

- the use of fly ashes, ground granulated blast-furnace slags and silica fume in concrete production (one of the best example of economical re-use of otherwise potentially health-injuring materials) is increasing;

- reuse of concrete for concrete production (a recycling, not a down-cycling process which has required the development of improved techniques and specific admixtures) is also expanding.

The main objectives of the environmental ready mixed concrete industry for the future remain the optimal consumption of materials, the reduction of outlets from production of raw materials and of concrete, and the re-use of the rest of the materials. Concern has been expressed, however, by the industry about the fact that, due to claimed limitation in duration and volume of production, other methods of concrete production, particularly site mixed concrete, are assumed somehow to be outside of strong environmental control.

The increasing complexity of environmental issues and pressure from competitors, customers and governmental organisations is forcing the industry to formalise its environmental management. This is the reason why, in the near future, a wider introduction of environmental Management Systems can be foreseen. To be really productive, these systems should be integrated with Quality Systems, in a way that tends to demonstrate both good production and environmental management to the internal and external interested parties.

THE IMPACT OF INFORMATION TECHNOLOGY

As well as other industries, the ready mixed concrete industry has been revolutionised in the last two decades by IT revolution, i.e., the introduction of electro-mechanically controlled equipment before, and of computer systems later.

Given the increased use of new materials, the capability of management of databases for constituent materials, as well as the thorough control of the production process of selected concrete, are of utmost importance,. These are the reasons why computer programs are currently in use in ready mixed concrete plants.

Computer programs are normally used to make comparisons and to investigate new materials, examining effects of modifying their proportions and comparing different production sources. Their use has at least, if not replaced, drastically reduced the number of laboratory trials required to establish, store and easily update basic production relationships: water demand and aggregate content vs. cement content, 7/28 concrete strength vs. water/cement ratio etc.

Sophisticated computer models have also been developed for simulation and enhanced design of concrete mixes. The most promising ones seem to be those based on the concept of particle interference, which progressively combine different materials, starting from the finest ones and taking into account their void filling capabilities and the effects of size [8].

A plant not equipped with extensive computer-based technology, which controls the overall production, can hardly be conceived today.

Control of mixing cycles and of weighing equipment, tracking of material consumption and storage management for accounting purposes, truck mixer control and customer identification are but some examples of tasks at present devoted to specific computer programs.

In this respect, what can be expected in the near future? As in almost all industrial sectors, strong impact of integration between information and communication technologies is highly probable, but interesting specific consequences can be envisaged. for the ready mixed concrete industry.

Even if it's evident that truck mixers' on-board radios are going to be substituted from more flexible and economic cellular phones, it should be realised that this could be the starting point of a revolution. Low cost GPS apparatus with the capability of transmitting data using cellular phone networks are entering the market. These integrated systems can not only control the location of truck mixers, thus enabling better logistics, but also drive actuators located on board. Such integration can therefore enable extensive control of the activity of the mixer– drum revolving, charge and discharge, etc. in such a way that the truck mixer itself could be considered a real remote controlled part of the production plant.

This means, for example, that critical tasks of the production process, like the addition of water and admixtures for the adjustment of concrete consistence at site, could be safely monitored from remote locations, instead of being under the responsibility of the truck driver.

The development of cellular communication will also give new impulse to technologies, like concrete production in unattended remote controlled plants, which, after some attempts in the late 70s, came to an early standstill because of the difficulty in operating a non continuous production in an economic way using dedicated cabled based networks.

MATERIALS AND PRODUCTION

The twentieth century "...will be remembered for the significant advancement in the science and technology of materials. Although concrete is a product of simple technology, it possesses a complex microstructure which is sensitive not only to the quality of component materials and mix proportions, but also to the processing methods used in construction practice..." [9]. These are the reasons why, in the case of ready mixed concrete, materials and production processes are so linked together.

Besides different types of cement, the concrete industry has gained vast experience in the use of different types of materials in an economical way. Designing materials to enable the best use of concrete, and designing concrete to make the best use of available materials, are commonly identified as the basis of any successful concrete production [10]. The concrete industry today can either utilise ready to use products or achieve equivalent performance by using separate components in the batching process. Complexity is no longer a problem, provided that optimised, computerised methods are used.

Between the different materials, special mention has to be made of additions and admixtures. In the two last decades their application has undoubtedly led to technical, economical and ecological advantages which have strongly improved the overall performance of concrete.

The concrete industry, due to its size, is a natural candidate for the disposal of industrial by products. The use of supplementary cementitious materials like slag, fly ash and silica fume as additions in concrete is one of the best example of economical and ecological reuse of otherwise environmental critical industrial waste. Reducing concrete permeability and enhancing the resistance to embedded steel corrosion, freeze-thawing, alkali-silica reaction and sulphate attack, additions support the effort of the industry aimed to promote the use of concrete as an economic but durable material. "In the industrial world it will be hard to find similar examples of excellent complementary or marriage between two components of a system, one of which happens to be an industrial waste"[9].

The improved use of admixtures has stimulated the use of ready mixed concrete for new challenges in applications: Self Levelling Concrete and Sprayed Concrete are but some examples of special concrete which could not be economically, and sometimes also technically, produced by the industry without the use of proper admixtures. The admixture industry is also proposing economical solutions for the reuse and recycling of fresh returned concrete and for minimising the need to dispose of high alkaline wash-water wastes. Strength evolution in time still remains the main item to assess the performance of almost all the concrete admixtures, but other performance related parameters like setting process and air content, are widely used in the new European performance-oriented admixture standard [11].

NEW CHALLENGES FOR THE 21st CENTURY RM CONCRETE INDUSTRY: FROM CONCRETE PLANTS TO CONCRETE FACTORIES

Besides strength, both in small and large projects, durability is more and more accepted as a basic requirement for concrete structures: structures should last for their service life without requiring heavy expenditure for repair. Even if this goal has been unanimously recognised and is now considered in almost all design and material standards, other concrete properties are of utmost importance for the construction industry: slump retention, good workability combined with absence of bleeding and segregation, low shrinkage and creep.

In a site cast concrete building frame today, labour costs (form preparation, reinforcing, concrete, form dismantling) almost equal the material costs. This is the reason why, for an increased use of site cast concrete, attention has to be paid to techniques and solutions that can reduce the manpower involved. A possible solution is the so-called "self compacting concrete", i.e. concrete which, without segregation, has great mobility under small activating forces. Such concrete is now produced by many ready mixed concrete plants and it is extending the market share in many countries.

The use of High Performance Concrete (HPC), with strengths up and above 100 N/mm^2, is also expanding, especially in Nordic Countries. In order to satisfy not only enhanced strength, but also increased durability demand, HPC is used in bridges, high-rise buildings, off-shore and marine structures, pre-stressed elements. The challenge for the ready mixed concrete industry is to manufacture and deliver such concrete with locally available materials and the proper use of additions, and of new generation admixtures.

New niche markets and other challenging opportunities are also foreseeable: polymer modified concrete and fibre reinforced concrete are but examples of what applied research has proposed in recent past to the ready mixed concrete producers and to the building industry.

New products and new challenges somehow require new industrial organisation. This is also the case for the 21st century ready mixed concrete industry. From a production point of view, a ready mixed concrete plant today still depends largely on the individual capabilities of the workers that operate in it. This should no longer be the case in the future as resource optimisation makes it necessary to pay maximum attention not only to the "internal" production factors (raw materials, personnel, equipment), but also to the so-called "external" factors. These include all resources that are in any way effected by the production process.

Therefore, a ready mixed concrete plant will be more and more organised in an industrial way, with the aim to optimise all human and material resources and to guarantee a constant product quality.

Such an industrialised production process will integrate the individual skills of workers in an organisational frame-work which reduces the possibilities for human error. Dividing the work into specialised functions, and focusing the role of the workers on planning, managing and controlling the computer-driven production process, means that in the near future the concrete industry will move far away from ready mixed "concrete plants" towards ready mixed "concrete factories".

CONCLUSIONS

For at least 6000 years, concrete has dominated the scene of the construction industry due mostly to its easy availability, versatility and low cost. In this century, "ready mixed" concrete, produced and supplied ready-off-the-shelf in an industrialised way, has become more and more recognised as an efficient and cost effective solution for the execution of concrete works.

It is obviously of primary interest for the ready mixed concrete industry to promote the use of concrete in the different sectors of the building industry. To increase the concrete market's share against other building materials in civil works for infrastructure networks, as well as in industrial and residential buildings, new challenges, both from the technological and environmental side, have to be faced:

- new products and materials
- improved production systems and organisational issues
- implementation of sound technical standards and respect for the environmental legislation.

As P K Mehta wrote, "...(in the twenty-first century) the future of the cement and concrete industries would very much depend on our ability to inter-link the growth of these industries with the goal of sustainable development..."[9]. No doubt the ready mixed concrete industry is prepared to do all its best for concrete to continue to be considered as the optimal construction material, and ready mixed the best way to supply it.

REFERENCES

1. VITRUVIUS, P. "De Architectura" Book II Chap. IV. Trans. by J Gwilt, London, 1826.

2. BONNER, M. The Origins of Ready Mixed Concrete. Presentation given at the ERMCO Annual Meeting of Representatives, Dresden, 1996.

3. DEWAR, J D AND ANDERSON, R. Manual of Ready Mixed Concrete, 2nd Edition, Blackie Academic and Professional, 1992.

4. COUNCIL DIRECTIVE 89/106/CE on the Approximation of the Laws, Regulations and Administrative Provisions in the Member States Relating to Construction Products. Off. Jour. of European Communities C136 – 4.6.1985.

5. EN206 Concrete Performance. Production and Conformity (Draft 24), Brussels, 1999.

6. QSRMC Manual of Quality Systems for Concrete, The Quality Scheme for Ready Mixed Concrete, Hampton, 1986.

7. Council Directive 98/C333/02 on Landfill of Waste. Off. Jour. of European Communities, 30.10.1988.

8. DEWAR, J D. Development of a Theory of Article Mixtures and its Application to Aggregates, Mortars and Concretes. PhD Thesis, submitted December 1997, City University, London.

9. MEHTA, P K. Bringing the Concrete Industry into a New Era of Sustainable Development. Mario Collepardi Symposium on Advances in Concrete Science and Technology. Fifth CANMET/ACI International Conference Proceedings, Rome, 1997.

10. NEVILLE, A M AND BROOKS, J J. Concrete Technology. Longmans Scientific and Technical, 1994.

11. EN934-2 Admixtures for Concrete, Mortars and Grouts, Part 2, Brussels, 1998.

INTERFACING INNOVATION WITH BEST PRACTICE

P C Hewlett

British Board of Agrément

United Kingdom

ABSTRACT. Product quality and client satisfaction governs success. Change and innovation are perceived as carrying risk and their adoption is usually approached with caution. Nevertheless, innovation within the construction and building sector is essential if the business activity is to progress and benefit from new ideas and competition.

However, innovation based on research and development is not supported as an activity within this sector as in other industrial sectors such as electronics and pharmaceuticals. The reasons for this disparity are discussed and the relationship between innovation and invention suggested and the distinction between Technical Approval and Standardisation made.

Those factors that may encourage or hinder innovation are identified and the roles of regulation and certification/performance assessment are explored since the adoption of innovation into reliable and best practice will govern whether commercial benefit and client well being are the ultimate results.

Keywords: Innovation, Quality, Performance, Testing, Research, Standards, Approvals, Regulation.

Professor Peter C Hewlett is the Director of the British Board of Agrément and Visiting Industrial Professor to the Department of Civil Engineering at the University of Dundee.

He specialises in materials for construction and building both inorganic and organic and in particular the enhanced performance of concrete by way of chemical modifications.

He is President of the European Organisation for Technical Approvals and the UK Concrete Society.

INTRODUCTION

For a progressive future change is inevitable. This truism applies to all matters and construction and building products and systems are no exception. To change we have to innovate but like so many commonly used words their meaning becomes blurred. Therefore I will start with a definition.

Innovation is the adverb derived from the verb to innovate and both relate to the noun innovation. Since we are concerned with building products, the noun is probably more appropriate and may be defined as alteration, change, new idea, departure, introduction, newness, novelty and variation.

Innovation is above invention since it is more deliberate and less random being based upon knowledge rather than flair. Additionally, innovation does not stop at the product itself but is also concerned with the adoptive step of incorporating a product or process into practice. In this regard invention is part of the innovation process.

Two impressions are obvious, namely that innovation is dynamic and results in change and secondly, departure and variation are recognised. This is sometimes referred to as deviation meaning departure from what is regarded as normal or in the case of products, as conventional and established. Because innovation is dynamic it sometimes creates uncertainty and carries with it a perceived risk. Indeed, lack of innovation may well reflect a lack of personal and/or national confidence.

A number of factors work against accepting innovative products. For instance,

- the product is harder to sell
- indifference to benefits of change
- ignorance
- attachment to old ideas
- fear of remedial costs
- risk and assumed liability
- lack of confidence

Problems of accommodating innovation are global and in the construction and building sector, some would argue against innovation. However, and by way of example in the USA, the value of buildings and infrastructure is put at $20 trillion ($20^9$) and getting the best from such investment would seem to make sense [1]. The objective of innovating is to improve the product and process of construction.

Some industries rely on innovation to survive. For instance, pharmaceuticals, electronics, aeronautical engineering and plastics technologies. However the building and construction sector accept change only gradually and it is suspicious of new things.

Influencing Factors

Within the field of concrete there have been relatively few fundamental innovations, e.g. inclusion of reinforcement as rods, bars or fibres, alternative hydraulic binders to lime and then Portland type cements although there have been numerous variations on established themes.

This attitude has its roots in the permanency and empiricism associated with buildings and constructions and these related activities. Over many years confidence and familiarity have been built up with certain types of material. For instance, brick, wood, iron and masonry. The alternatives of concrete, plastics and steel do not have the same pedigree of age although they have become accepted because they solved a real need and had measurable benefits.

It is perhaps more the current range of innovations such as coatings, new generation composite plastics, alloys and modified cements that raise questions and doubts.

In this regard many professions are involved in helping or hindering innovation and its adoption. For instance, corporate executives, designers, project managers, craftsmen, legislators, insurance underwriters, government departments/officers, regulators, lawyers, inventors, manufacturers and contractors [1]. This is a very disparate collection of disciplines with different motivations that may help or hinder innovation.

However, there is only one creator of materials wealth and that is industry. Construction and building product manufacture are part of industry and hence wealth creation. If new products and systems are not accepted and brought into active use then building practice and designs would stagnate.

It is usually intended that most innovations should be changes for the better, more efficient, longer lasting, lighter, cheaper or less dangerous. In other words, innovation helps to achieve functional improvement. Such a drive cannot be resisted and in that regard innovation is inevitable and has to be accommodated.

STAGES IN INNOVATION

There are four readily recognisable stages in innovation

1. Research or discovery

2. Development

3. Production and application

4. Exploitation/commercialisation

Basic research produces the majority of discoveries from which all other progress flows [2][3]. The motives behind basic research are often curiosity and understanding. As such they may seem indulgent and a little out of fashion with the current preoccupation with focussing effort, managing programmes and meeting targets.

These aspects are more appropriate to development than research and within construction and concrete research and development, often become synonymous and result in reduced quality research and/or poorly disciplined and irresolute development.

In this regard innovation is contemplative rather than haphazard. The drive is not altruistic but the creation of wealth and material wellbeing. This sequence is shown in Figure 1.

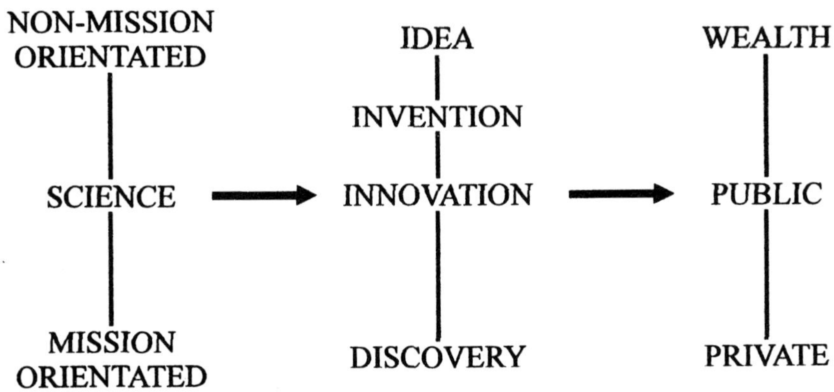

Figure 1 Innovation and wealth creation

There are levels of innovation, some minor, some not. For instance, a new concept might be regarded as having more potential importance than a new way of doing the same thing. The principle of using a high modulus reinforcement in a matrix having low tensile strength such as the inclusion of steel in concrete was at the turn of the century a new concept, This concept opened up design opportunities and there are now many variations on the same basic concept but these changes cannot compare with the initial idea.

According to Bernstein and Lemer [1] there have been 117 innovations in permanent residential structures over the period 1945 – 1990. [Table 1 – reproduced with kind permission of the authors].

Innovation would nevertheless appear to be regional reflecting different priorities, one country to another. This is evident from Table 2 also taken from reference [1] comparing USA, Europe and Japan. Such a situation is probably best since it addresses national priorities and needs but also reduces duplication and offers trade opportunity between different countries.

Transposing scientific and technological potential into viable innovations has become recognised in the last decade as an important step in the creation of wealth. Within the building and construction sectors in the UK at least, it is competing with the funding of research and development within the building and construction sectors. Perhaps to the disadvantage of the latter. Clearly a balance has to be struck.

Table 1 Sample of innovations in permanent residential structures 1945 – 1990 [1]

FUNCTIONAL AREA	NO. OF INNOVATIONS
Structural exterior wall framing	07
Enclosure and insulation	08
Openings	13
Interior wall framing	07
Foundation	12
Floor framing	10
Roof framing	07
Roof covering	07
Plumbing	12
Electrical wiring	04
Heating/ventilation/air conditioning	12
Interior finish	18
Total	117

Table 2 Regional variability of innovation[1]

US LEADS	EUROPE LEADS	JAPAN LEADS
High-performance concrete	High-performance asphalt	High-performance steel
Waste/wastewater treatment	High-speed rail/Magnetic levitation vehicles (Mag-lev)	Automated equipment
Computer Aided Design (CAD)/ Computer Aided Engineering (CAE)	Tunnelling	Field computer use
Solid/hazardous waste disposal	Real-time site positioning systems	High-speed pavement assessment
Environment	Restoration	Safety
Geographic Information and Positioning Systems (GIS/GPS)	Marine construction	Intelligent Buildings
Integrated database	Energy conservation	Building systems

Relative levels of technology adoption in design and construction industries, as assessed in reconnaissance trips sponsored by the US National Science Foundation and the Civil Engineering Research Foundation. Source: Civil Engineering Research Foundation European Research Task Force. *Constructed Civil Infrastructure Systems R&D: A European Perspective, Washington, DC: Civil Engineering Research Foundation, 1994, p5*

Innovation and Research and Development

Research and development in the construction and building sectors is not a substantial percentage of total market value. Various estimates put the figure within the range 0.1 – 0.5% of turnover. Some sectors are above this level and others below. Notwithstanding this low level of reinvestment, the industry is inventive and new ideas and ways of doing things often result from experience and commonsense rather than scientific understanding. On occasions this ‘pioneering’ approach is often found wanting when a new innovation allegedly fails. If the product or system has not been based on a clear technical pedigree it is often difficult to substantiate the role played by the invention or innovation and to give it a credible defence. Such a background reinforces the need for independent performance-based assessments to establish the fitness and worth of an innovation. That is the role of Technical Approval and Agrément. It is not really the role of product conformity or compliance to standards. National and international standards are usually based on accumulated experience and reflect what has been and is being done. In this respect they are archival but serve the useful purpose of establishing comparative if not absolute performance levels. Approvals on the other hand depend on performance-based bespoke assessments reflecting function rather than historical practice.The advantages of this approach are clear and there is a growing trend for standards to become more performance-based. If this trend is maintained then assessment and compliance will become coincident. However the present situation is far from that point.

It is a characteristic of the building and construction sectors that appreciation of the need for sound science is not so obvious as in other activities such as medicine, space exploration and electronics. The reasons for this are complex having their root in the education of civil engineers and architects and the mechanics of specifying for construction and building work. Prescription by way of codes and standards pervades the activity and such a situation is likely to continue. This being so, science and its outturn through innovation have to be assisted in the transposition from scientific concept to new practice. In this regard standards, approvals, conformity and quality management certification all have a part to play. These interactions are shown schematically in Figure 2.

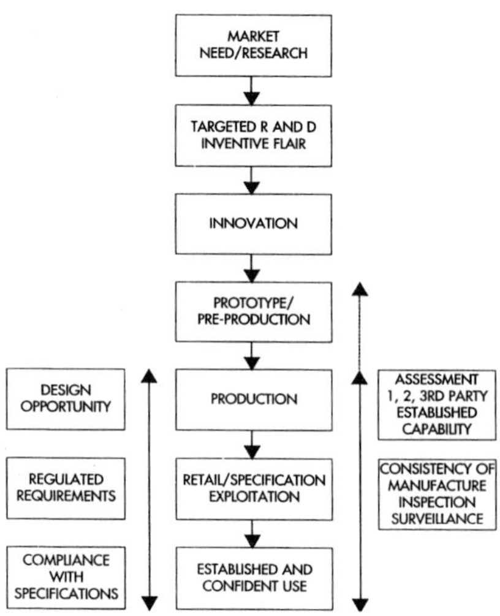

Figure 2 Transposing innovation into adopted practice

Not withstanding this situation there have been notable innovations in the construction and building sectors. For instance,

- high performance steels
- robotics
- high performance concrete
- high performance / engineered wood
- anti static flooring tiles
- silent pile driving equipment
- green technologies

In order to evaluate a new product it is important to have available test methods that can be used to measure the performance both in isolation and as incorporated in a building. Within the activity of standards making these are the so-called horizontal standards and are dealt with in the next section.

Test Methods

Many test methods are simulations of what happens in practice. They are useful for making comparisons, one product or system to another, but often do not permit cause and effect judgements to be made. For instance one may measure the vapour permeability of a protective membrane and it is a property of the material. Alternatively one may measure vapour transmission under a set of fixed conditions. Changing the conditions changes the value. Relating the value to a real event can be difficult if there is no absolute basis.

Much standardisation and certification work is product oriented with test procedures based on past experience with particular product types. In addition the measured responses also reflect the product types and not the actual performance required. For instance it is feasible to test an insulation in such a way as to reflect let us say, its fibrous nature, and to give ranges of values for thermal conductivity and transmittance that would typify such material. These values can be built into specifications so that only those materials that conform to such values would be chosen and used. This approach creates problems for new products that may impart their insulation properties (staying with the example) in quite a different way.

An alternative approach is to put emphasis on the performance required by way of characteristics and relate these to a measurement/test methodology and recording how the product responds. The designer can then used these measured characteristics in whatever way is appropriate. This approach can accommodate innovation.

Whatever approach is used they both depend upon having available agreed test methods. To design such tests it is necessary to understand the physics and perhaps the chemistry of the performance characteristics, e.g. freeze-thaw and durability testing of concrete.

The scale of test is another aspect that has to be considered. The interrelationship between testing a product, for instance a window, in the laboratory and relating that to the performance of the same window when installed in a brick wall, fitted with insulation and consisting of aerated concrete blocks for the inner leaf and a brick wall outer leaf can be very complex. Full scale testing is very desirable. For instance, windows and doors can be examined using the guarded hot box [5] [Figure 3]. Similar scaled up testing of roofing and external wall insulation can be assessed using BBA's Tempest rig [6] [Figure 4].

Figure 3 Guarded Hot Box apparatus

The BBA Guarded Hot Box is accredited for the determination of the thermal transmittance and conductance of building elements up to a size of 2.3m x 2.8m. It can also be used to test glazing and complex wall panels. The two main components of the apparatus, the hot chamber and the cold chamber, are mounted together in a frame which allows them to be rotated to change their orientation and permit the testing of horizontal elements such as ceiling panels or rooflights.

The use of a large number of calibrated thermocouples with computerised date acquisition enables the identification of cold bridges and the estimation of the potential condensation risk.

The BBA Tempest Rig comprises a computer that runs from specially software, a valve, an electrically drive radial fan, connecting flexible trunking and an enclosure (to suit the sample under test). The rig allows large building components to be subjected to simulated wind loading. A typical sample size is 3 x 3 metres. This apparatus has been recently adapted so that other tests may be run on external wall insulation systems using a vertical enclosure that can beclamped to the insulation system installed on a masonry wall.

These tests are described in EOTA (European Organisation for Technical Approvals) draft ETAG [7].

Figure 4 BBA Tempest Wind Loading apparatus

Such equipments are expensive representing considerable capital investment. In this respect some form of national facility having a greater chance of continuous use is to be preferred.

There is scope for international co-operation in the field of special facility provision and testing. This already exists on a limited basis within the UEAtc (European Union of Agrément)

Innovation and Regulation

Regulation generally and building regulations in particular provide a tempering of commercial enthusiasm that may otherwise only be concerned with profit, that on occasion may lead to excesses resulting in the health and safety of the population at large being put at risk. Regulation these days is also concerned with quality, fairness and the expectations of the purchaser/user whilst encouraging the manufacturers, suppliers/contractors to adopt good practices.

It is perhaps easier to realise such objectives by adopting well established routines and specifying products of good pedigree. Such an approach is 'safe' but acts against the use of new products and systems that may offer benefit.

Compliance with a national standard, preferably by a third party/independent accredited body is one way but if the variant product cannot comply with the recognised specification some alternative has to be available. That alternative is Agrément or national/European Technical Approvals.

The concept of approvals has now become worldwide with developing reciprocal arrangements coming into play between the national approvals bodies. The WFTAO (World Federation of Technical Approval Organisations) oversees such global cooperation. Very relevant if one exports as well as trading nationally.

In the UK the link between regulated requirements and national standards/Agrément is via what we call the Approved Documents. Recognition is given to approvals on a deemed to satisfy basis. However, there is a third option of an equivalent alternative. It is relevant to note that mutual recognition between various countries on compliance with Building Regulations has not developed very far. Harmonisation of regulations, even within Europe is more a matter of discussion that practice. Present attempts to evolve performance-based regulations (CIB TG 11) [8] could well ease if not entirely solve this problem.

At the moment Articles 16 and 17 of the CPD [4] require that country of destination requirements can be addressed in the country of manufacture but do not go so far as to accept the role of mutual recognition, as far as regulated requirements are concerned. This omission shows
itself from time to time. For instance, a manufacturer with a CE marked product will deem to satisfy regulated requirements in all member states of the EU*. The specification (with attestation) against which the CE mark is given, be it on the harmonised standards or ETAs, should therefore meet the regulated requirements in all member states. To achieve this tends to add requirements to the specification resulting in a 'maximalist' approach that becomes progressively more expensive. The alternative is to have a 'minimalist' approach keeping costs of compliance low and also not increasing the regulated level in any member country. This is the lowest common factor approach but has the disadvantage of not necessarily satisfying the needs in some member states, that in turn can lead to arbitrary additional requirements resulting in possible technical barriers to trade. The very objective the CPD was designed to prevent.

Innovative products, it may be argued should be subject to more scrutiny than those that are established and can be covered by standards. There is some truth in this view and would justify a more maximalist approach to approvals. If that was the agreed position it would have to be accepted by all parties and regarded as appropriate by manufacturers and their collective trade body representatives so that the technical underwriting of innovative products by way of approvals, at whatever level, was accepted.

The support for technical approvals should be in proportion to that for standards making if innovation is to be encouraged and exploited.

Regulators and specifiers should encourage responsible innovation and therefore also encourage the use and development of Technical Approvals and Agrément.

A balance has to be struck between caution where there is risk and yet recognising the continuum between idea/innovation, approvals, regulation and ultimately standardisation. This sequence is shown in Figure 5[9].

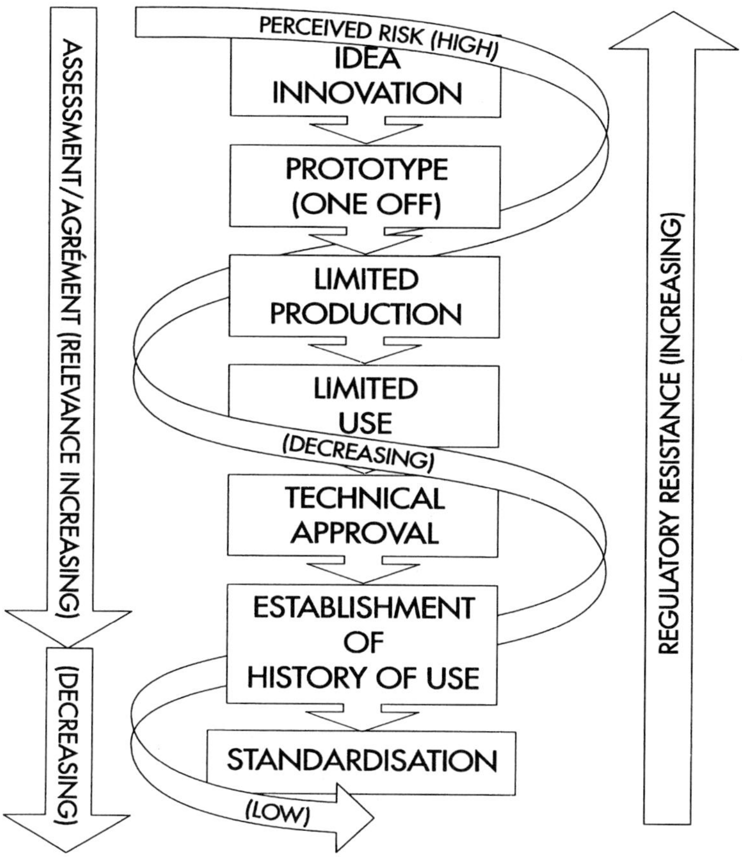

Figure 5 Perception of risk from innovation to established use

Innovation is both inevitable and desirable. It should be encouraged and the products of innovation used so long as their fitness for purpose has been established.

Innovation has its roots in the science of building pathology and functional performance. Only if we understand can we innovate in an objective and logical way. Innovation can, however, be confused with invention. The latter is unpredictable and haphazard originating from flair and practical awareness rather than scientific understanding. However, in the building and construction sectors invention can be significant. Even so, the results of invention should still be subject to rigorous performance assessment.

At the present time there are no clear and agreed criteria by which organisations may be recognised as competent to compile a performance-based assessment specification and deliver the approval against it. There are however such criteria for approved testing, inspection and certification bodies at least in the European Union [10]. This situation may have to be considered during the next decade.

CONCLUSIONS

1. Innovative building products are essential for a vigorous and competent building and construction activity. They should not be regarded with suspicion and considered second best if fitness for purpose has been established by a credible third party assessment/approvals body.

2. Many countries are fortunate in having this service available to its product manufacturing industry or if not already established has the latent capability to create it. This service should be used to the full ultimately carrying forward and helping to create better national and international standards. It may also offer a prospect in demonstrating regulatory compliance one country to another.

3. Only in this way will innovative concrete products be readily used justifying the expenditure on the research and development to begin with.

REFERENCES

1. BERNSTEIN, H M and LEMER, A C. "Solving the Innovation Puzzle" ASCE Press, New York 1996 pp 127

2. BULLETIN OF THE EUROPEAN UNION SUPPLEMENT. "Green Paper on Innovation" 5/95, European Commission, 1996, pp 102

3. LANGRION, J , GIBBONS M , EVANS W G and JERONS, F R. "Wealth from knowledge' a study of innovation in industry, Macmillan 1972, pp 477

4. The Construction Products Directive 89/106/EEC, Official Journal of the European Communities, 21 December 1988

5. BRITISH STANDARDS INSTITUTION British Standard 874: Part 3: Section 3.1, 1987,

6. BRITISH BOARD OF AGRÉMENT:
   BBA Method Of Assessment And Testing, No 50, 'UEAtc Technical Guideline for the Assessment of thermal insulation systems intended for supporting waterproof coverings on flat and sloping roofs'. November 1992,
   BBA MOAT No 55, Supplementary Guide for the assessment of mechanically fastened roof waterproofing systems' April 1991

7. EUROPEAN TECHNICAL APPROVAL GUIDELINE 'Guideline for Technical Approval for external thermal insulation composite systems with rendering', private communication EOTA (not published)

8. CIB TG11 'Performance-based building codes' May 1998. Prepared by The Inter-jurisdictional Regulatory Collaboration Committee.

9. HEWLETT, P C and DE NORMANN, J. 'Innovation and related matters: Technical Approvals and Standards'. BSI Committee B/-/2 Working Group report (unpublished document) August 1995

10. EUROPEAN STANDARDS ORGANISATION:
EN45001: 1989 "General criteria for the operation of testing laboratories"
EN45004: 1995 "General criteria for the operation of various types of bodies performing inspection"
EN45011: 1989 "General criteria for certification bodies operating product certification"
EN45012: 1989 "General criteria for certification bodies operating quality system certification"

EUROCODE 2 - INNOVATIONS IN DESIGN AND CONSTRUCTION OF CONCRETE STRUCTURES

H-U Litzner

German Concrete Society

Germany

ABSTRACT. The Structural Eurocodes of the European Communities establish requirements for building and civil engineering works in terms of reliability, adequate performance in service conditions and durability. For the achievement of durability, several steps are necessary in the design process. They are subject of Eurocode 2 and the European Standard EN 206 for concrete technology. The basic elements of this integrated design concept are described in the present paper.

Keywords: European Standards for the Design of Concrete Structures, Design Working Life, Durability, Environmental Actions, Concrete Technology, Cover to Reinforcement, Crack control, Execution, Curing of concrete.

Dr H-U Litzner is managing director of the German Concrete Society. He specialises in design of concrete structures and concrete technology. Since 1980, he is involved in the Eurocode programme. Since 1990, he is Chairman of CEN/TC250/SC2 which is responsible for Eurocode 2 "Design of Concrete Structures".

STRUCTURAL EUROCODES AND THEIR OBJECTIVES

For the realisation of the European single market, the Commission of the European Communities (CEC) has initiated the work of establishing a set of unified technical rules for the design of building and civil engineering works which will gradually replace the different rules in force in the various EC-Member States. These technical rules which became known as the *Structural Eurocodes* shall lead to structures which fulfil the following fundamental requirements established in [1]:

1. With acceptable probability, they will remain fit for the use for which they are required, having due regard to their intended life and their cost.

2. With appropriate degrees of reliability, they will sustain all actions and influences likely to occur during execution and use and have adequate durability in relation to maintenance costs.

In other words, the fundamental requirements which shall be met are adequate *performance* in use, appropriate degree of *reliability* and adequate *durability* during the design working life (Table 1). The relationship between durability and *maintenance costs* should be noted.

The *Structural Eurocodes* provide the technical tools for the achievement of these requirements. The corresponding elements of the design concept are described in the following. They are related to Classes 3 and 4 in Table 1, where the design working life is defined as follows [1]:

"... The design working life is the assumed period for which a structure is to be used for its intended purpose with anticipated maintenance but without major repair being necessary."

Table 1 Indication of the design working life required in [1]

CLASS	REQUIRED DESIGN WORKING LIFE IN YEARS	EXAMPLE
1	1-5	Temporary structures
2	25	Replaceable structural parts, e.g. gantry girders, bearings
3	50	Building structures and other common structures
4	100	Monumental building structures, bridges and other civil engineering structures

EUROPEAN STANDARDS SYSTEM FOR CONCRETE STRUCTURES

Figure 1 presents the actual European Standards System for building and civil engineering works in concrete, which consists mainly of European Prestandards (ENV). They are actually converted to European Standards, which will replace the corresponding national standards in force in the various CE-Member States.

In this European Standards System which provides all elements for structural and durability design four levels can be distinguished:

- Level 1 comprises standards for structural safety [1] and actions on structures; in particular, in [1] basic durability requirements are established.
- Level 2 consists of Eurocode 2 [2] for the design and detailing of concrete structures.
- Level 3 gives data for structural materials, in particular for concrete [3], [4], and the execution of concrete structures [5].
- Level 4 consists of standards for the testing of materials.

It should be noted, however, that the standards shown in Figure 1 will only lead to the performance required in [1], [2], if they are applied in appropriate combination.

VERIFICATION OF THE DURABILITY OF CONCRETE STRUCTURES

According to [1], [2] it shall be verified that a concrete structure satisfies the following condition:

$$S_d \leq R_d \qquad (1)$$

where:

S_d denotes the design value of an action effect

R_d is the corresponding design resistance associating all structural properties with the respective design values.

Equation (1) has initially been derived for *direct* actions such as, for example, permanent or imposed loads, and for *indirect* actions such as, for example, imposed constraint or deformation (e.g. due to hydration heat).

In the context of [1], [2] and [4], the format given by Equation (1) can also be used for environmental actions ([6] and Figure. 2). For example, R_d can be interpreted as the actual concrete cover c and the corresponding design value S_d as the carbonation depth. This will be illustrated in a later section.

Such methods for durability design require, however, well defined performance characteristics, precise tests methods, reliable models for material behaviour and good knowledge of the environmental conditions. The values of S_d are strongly related to climatic parameters and to other deterioration factors as described later.

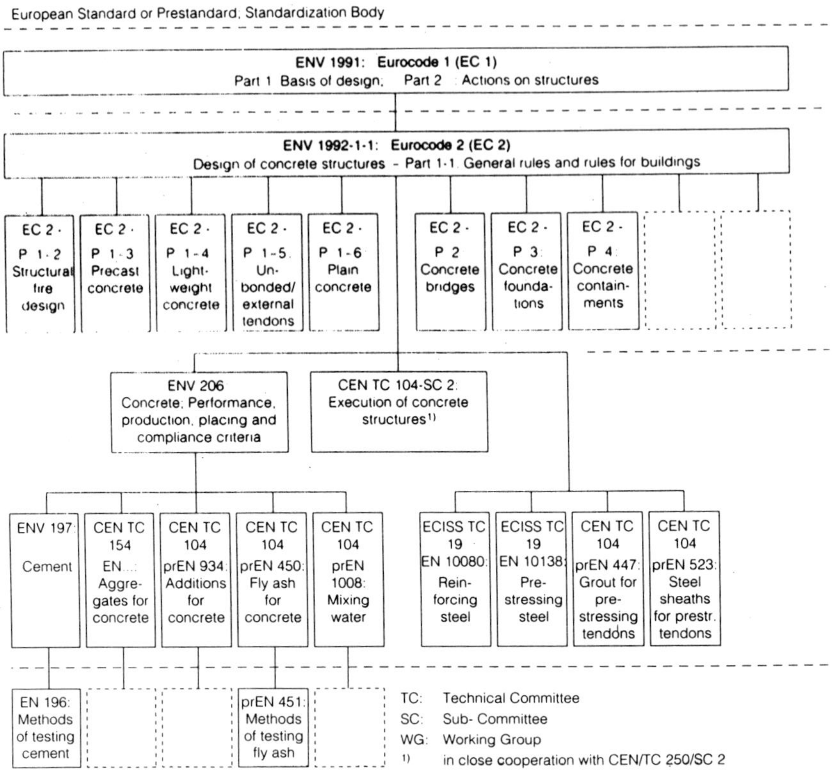

Figure 1 Structure of the actual European Standards System for building and civil engineering works in concrete

The design resistance R_d of concrete structures depends on several parameters. The most important of them are:

- Permeability and structure of the concrete
- Crack pattern and crack width
- Type of reinforcement (steel reinforcement, prestressing steel)
- Cover to reinforcement
- Quality of workmanship during execution (e.g. curing)

The possible influence of those parameters on R_d will be considered in the subsequent sections.

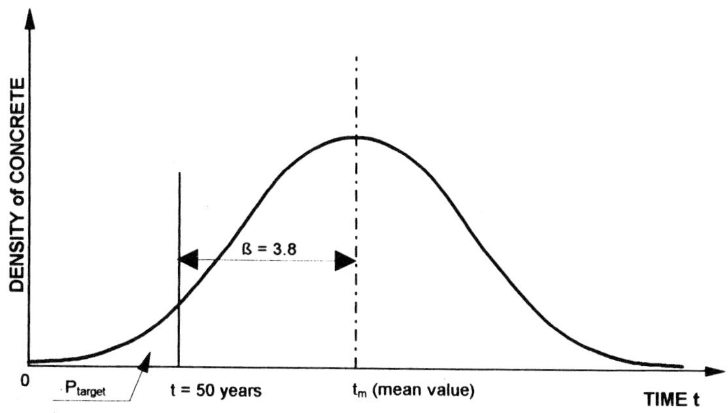

Figure 2 Example of a service life distribution with a reliability index $\beta = 3.8$ and an intended service life of $t = 50$ years [6]. The probability p_f of a failure (i.e. $S_d > R_d$) during the intended lifetime of 50 years shall be smaller than an assumed target probability ($p_{f,\,tar} = 7 \cdot 10^{-5}$ in [1])

DEFINITION OF THE ENVIRONMENTAL ACTIONS S_d

According to [1], a concrete structure shall be designed in such a way that deterioration of concrete and/or steel (see Figure 3) should not impair the durability and performance of the structure, having due regard to the anticipated level of maintenance. In other words, an adequate maintenance strategy is part of the design concept of the Structural Eurocodes.

The above requirements to be met by concrete structures depend mainly on the environment to which the concrete is exposed. Environment in this context implies chemical and physical actions resulting in effects (see Figure 3) which are not considered as loads in structural design.The environmental actions defined in [4] are shown in Tables 2 and 3 where rough distinction is made between six deterioration mechanisms for concrete and steel respectively.

The actions in Tables 2 and 3 may, where relevant, be considered as *local* or *micro* conditions. Local conditions are those around the structure after having been built, taking into account the specific actions where the structure or the structural element is located (e.g. realitve humidity RH, CO_2-content).

However, in some circumstances, micro conditions need to be considered. These denote environmental actions on a specific surface of a structural element. This may, for example, apply to the following circumstances:

- Exposition to driving rain
- Exposition to sun radiation
- Contact with earth, ground water, sea water etc.

Table 2 Exposure classes defined in [4]

DETERIORATION MECHANISM	CLASS DESIGNATION	DESCRIPTION OF THE ENVIRONMENT	INFORMATIVE EXAMPLES WHERE EXPOSURE CLASSES MAY OCCUR
1 No risk of corrosion or attack	XO	Very dry	Concrete inside buildings with very low humidity (RH < 45 %)
2 Steel corrosion induced by carbonation	XC1	Dry	Concrete inside buildings with low humidity (RH < 65 %)
	XC2	Wet, rarely dry	Parts of water retaining structures, many foundations
	XC3	Moderate humidity (RH < 80 %)	Concrete inside buildings with moderate or high air RH; external concrete sheltered from rain
	XC4	Cyclic wet and dry	Surfaces subject to water contact, not within class XC2
3 Steel corrosion induced by chlorides	XD1	Moderate humidity	Concrete surfaces exposed to direct spray containing chlorides
	XD2	Wet, rarely dry	Swimming pools; concrete exposed to industrial water containing chlorides
	XD3	Cyclic wet and dry	Parts of bridges; pavements; car park slabs
4 Steel corrosion induced by chlorides from sea water	XS1	Exposed to airborne salt, not in direct contact with sea water	Structures near to or on the coast
	XS2	Submerged	Parts of marine structures
	XS3	Tidal, splash and spray zones	Parts of marine structures
5 Freeze/thaw attack on concrete	XF1	Moderate water saturation, no de-icing agents	Vertical concrete surfaces exposed to rain and freezing
	XF2	Moderate water saturation, with de-icing agents	Vertical concrete surfaces of road structures exposed to freezing and airborne de-icing agents
	XF3	High water saturation, no de-icing agents	Horizontal concrete surfaces exposed to rain and freezing
	XF 4	High water saturation, with de-icing agents	Road and bridge decks exposed to de-icing agents and vertical concrete surfaces exposed to direct spray containing de-icing agents and freezing
6 Chemical attack on concrete	XA1, XA2, XA3	Aggressive chemical environment	See Table 3

It becomes evident that the approach in the design for durability is similar to structural design where normally *global* verification (e.g. structural analysis) and *local* checks (e.g. stress limitation) are performed.

Table 3 Limiting values for exposure classes XA for chemical attack in [4]

CHEMICAL CHARACTERISTIC	XA1	XA2	XA3	TEST METHOD
SO_4^2 mg/l in water	≥ 200 and ≤ 600	> 600 and ≤ 3000	> 3000 and ≤ 6000	EN 196-2[7]
SO_4^2 mg/kg in soil[1)] total amount	≥ 2000 and ≤ 3000[2)]	> 3000[2)] and ≤ 12000	> 12000 and ≤ 24000	EN 196-2[3)]
ph of water	≤ 6.5 and ≥ 5,5	< 5.5 and ≥ 4.5	< 4.5 and ≥ 4.0	DIN 4030-2 [8]
Acidity of soil	> 20 ° Baumann Gully			DIN 4030-2
CO_2 mg/l aggressive in water	≥ 15 and ≤ 40	> 40 and ≤ 100	> 100	
$_4NH^+$ mg/l in water	≥ 15 and ≤ 30	> 30 and ≤ 60	> 60 and ≤ 100	ISO 7150-1[9] ISO 7150-2[10]
Mg^{2+} mg/l in water	≥ 300 and ≤ 1000	> 1000 and ≤ 3000	< 3000	ISO 7980 [11]

Footnotes:

1. Clay soils with a permeability below 10^5 m/s may be moved into a lower class.
2. The 3000 mg/kg limit is reduced to 2000 mg/kg, where there is a risk of accumulation of sulphate ions in the concrete due to drying and wetting cycles or capillary suction.
3. The test method prescribes the extraction of SO_4^2 by hydrochloric acid; alternatively, water extraction may be used, if experience is available in the place of use of the concrete.

Figure 3 Deterioration of concrete due to environmental actions

RESISTANCE OF CONCRETE AGAINST ENVIRONMENTAL ACTIONS

EN 206 [4] provides two general methods for the assessment of the design resistance R_d in Equation (1). The standard method ("macro-level design" in [6]) consists of the provision of limiting values for concrete composition in terms of maximum water/cement-ratio, minimum cement or air content and, where relevant, of additional requirements for cement and/or aggregates (see Table 4). Alternatively, performance - related methods with respect to durability may be used. These may be based on refinements of the standard method, on approved and proven tests or on analytical models.

When using the latter approach, for example, for environmental class XC in Table 4, i.e. deterioration of steel by carbonation , it shall be verified that

$$d_c \leq c_{act}, \tag{2}$$

where d_c denotes the carbonation depth and c_{act} the actual cover to reinforcement of the member considered.

From Equation (3) it can be concluded that d_C - or, in more general terms, the physical model - depends on several parameters with a statistical distribution. These parameters may be characterized as follows:

- the standard deviation of their statistical distribution is high;
- they are significantly time - dependent;
- the „scale effect" should be considered, i.e. it should be checked whether results from lab tests can be applied in insitu conditions.

It should be noted, however, that the actual development in the field of durability is characterized by world-wide activities with the objective to define design values of all relevant parameters which can be introduced in the verification format described above (see Figure 5 and the references in [6]).

The carbonation depth d_c may be calculated from:

$$d_c = \sqrt{2 \cdot k_1 \cdot k_2 \cdot k_3 \cdot \Delta_{con}} \quad \sqrt{\frac{D_{nom}}{a}\, t} \quad \left(\frac{t_0}{t}\right)^n \leq c_{act} \tag{3}$$

where:

d_c the carbonation depth

D_{nom} the diffusion coefficient of dry concrete for carbon dioxide in defined environment (20°C, 65 % rel. humidity)

a the amount of CO_2 for complete carbonation

Δ_{con} the concentration difference of carbon dioxide at the carbonation front and in the air

k_1, k_2, k_3 parameters for micro climatic conditions, to describe curing conditions and the effect of water separation (local w/c-ratio) respectively

n parameter for micro climatic conditions, describing wetting and drying

$n = 0$ for interior conditions; $n \leq 0.3$ for outdoor conditions

t_0 reference period, $\sqrt{t}$ -law valid (e.g. 1 year)

t time

c_{act} actual cover to reinforcement

Table 4 Limiting values for composition and properties of concrete made with cement type CEM I conforming to ENV 1997-1 [12]

EXPOSURE CLASS	LIMITING VALUES FOR CONCRETE COMPOSITION			
	w/c-Ratio	Strength Class	Cement Content kg/m^3	Air Content [%]
XO	—	C12/15	—	—
XC1	0,65	C20/25	260	—
XC2	0,60	C25/30	280	—
XC3	0,55	C30/37	280	—
XC4	0,50	C30/37	300	—
XS1	0,50	C30/37	300	—
XS2	0,45	C35/45	320	—
XS3	0,45	C35/45	340	—
XD1	0,55	C30/37	300	—
XD2	0,55	C30/37	300	—
XD3	0,45	C35/45	320	—
XF1	0,55	C30/37	300	— 1),2)
XF2	0,55	C25/30	300	4,0 1),2)
XF3	0,50	C30/37	320	4,0 1),2)
XF4	0,45	C30/37	340	4,0 1),2)
XA1	0,55	C30/37	300	—
XA2	0,50	C30/37	320	— 2)
XA3	0,45	C35/45	360	— 2)

1) freeze / thaw resisting aggregates

2) Sulfate resisting cement

CONCRETE COVER TO REINFORCEMENT

According to Eurocode 2 [2], a *nominal* concrete cover to reinforcement shall be introduced in the design calculations. It is given by:

$$\text{nom c} = \text{min c} + \Delta h \qquad (4)$$

where:

nom c	denotes the nominal cover
min c	is the minimum cover
Δh	is an allowance for tolerances.

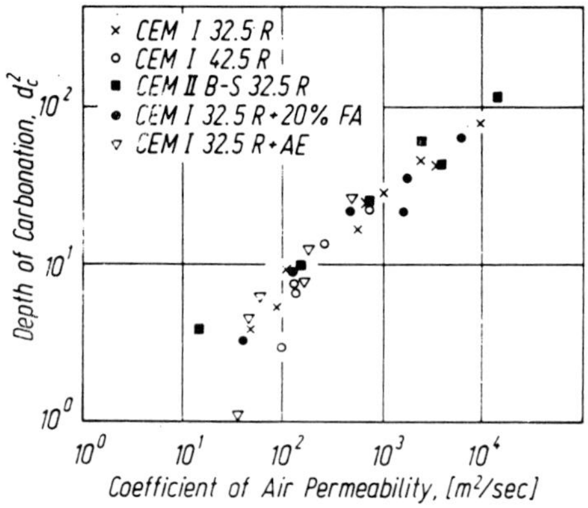

Figure 4 Depth of carbonation of concrete after 1 year of storage at 20°C, 65 % RH, as a function of the air permeability coefficient at an age of 56 days for concretes made of various types of cements [13]

For the determination of the *minimum* concrete cover, *min c*, the following criteria apply:

- Safe transmission of bond forces
- Avoidance of spalling
- Adequate fire resistance
- The protection of the steel against corrosion

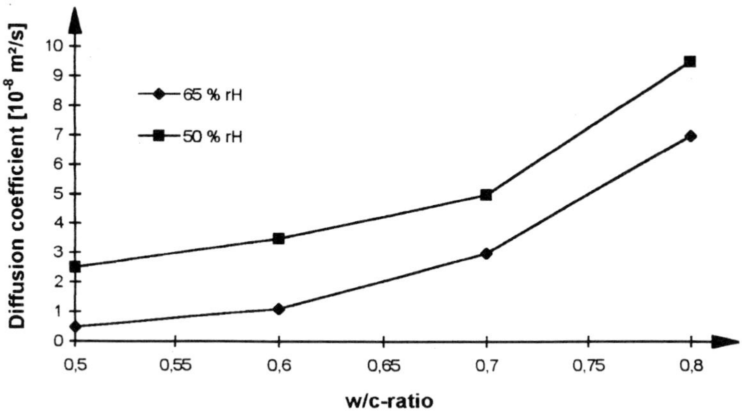

Figure 5 Example for the diffusion coefficient of concrete as function of the w/c-ratio for two relative humidities

In the latter case, the protection against corrosion depends upon the continuing presence of a surrounding alkaline environment provided by an adequate thickness of good quality, well-cured concrete (see later section). In the absence of other provisions, adequate thickness may be assumed if the values of *min c* given in Table 5 are used [15]. Except for exposure class XC1 they may be reduced by 5 mm for slab elements.

A further point concerns the trade-off between *min c* and the actual concrete grade used (see Table 6). Generally, a reduction of the values in Table 5 is allowed provided that the actual concrete grade is higher than the minimum grade given in Table 4 for the relevant exposure class. For example, in [15] which is based on [2], the trade-off is restricted to prefabricated concrete elements and a reduction of 5 mm of the values in Table 5 is permitted for normal strength concrete ($f_{ck,cylinder} \leq 55$ N/mm^2) in the following cases:

- Exposure Class XC1 : no reduction permitted
- Classes XC2, XC3 : reduction permitted if the actual concrete grade corresponds to the minimum strength class in Table 4
- Class XC4 : reduction permitted for C35/45 and higher
- Classes XS, XD : reduction permitted if the concrete grade is at least two strength classes higher (see Table 6) than the minimum strength class in Table 4

Table 5 Minimum cover, *minc*, to reinforcement for corrosion protection [mm]

TYPE OF STEEL	EXPOSURE CLASS IN RELATION TO STEEL CORROSION									
	XC1	XC2	XC3	XC4	XS1	XS2	XS3	XD1	XD2	XD3
Reinforcing steel	10	20	20	25	40	40	40	40	40	40
Prestressing steel	20	35	35	35	50	50	50	50	50	50

For high strength concrete (see Table 6), additional consideration may be necessary. The allowance for tolerances, Δh, will usually be in the range of 0 mm < Δh < 5 mm for precast concrete elements if the production control can guarantee these values. The allowance will be in the range of 5 mm < Δh < 10 mm for in-situ reinforced concrete construction. Higher or lower tolerances, Δh, may be used if this can be justified by the construction method used. These rules underline that for durability design a close relationship between concrete technology and construction on site exists.

CONTROL OF CRACKING

The durability of concrete structures may adversely be affected by excessive cracking. Besides that, cracking shall be limited to a level that will not impair the proper functioning of the structure or cause its appearance to be unacceptable.

Cracks in concrete structures have two primary causes:

- Cracks caused by the rheological properties of the fresh or hardening concrete
- Cracks caused by loading and/or imposed deformations

The first type of cracks can be controlled by appropriate measures of concrete technology, in particular by the composition of the concrete mix, proper placing and curing. Corresponding rules are provided in [3 to 5]

For the control of cracks caused by loading and/or imposed deformation, the design concept in Eurocode 2 provides two basic tools:

- The requirement of a minimum bonded steel reinforcement
- The limitation of crack width

The minimum steel reinforcement has two functions: it should ensure an equilibrium at the time when cracks may first be expected. In addition, the area of the minimum reinforcement should be such that crack widths with an unacceptable value are avoided. In most cases, the minimum reinforcement is calculated for imposed deformations due to the dissipation of the hydration heat, i.e. for a concrete age between 3 to 5 days after casting. It depends mainly on the actual concrete tensile strength, f_{ct}

Table 6 Strength classes for normal-weight and heavy-weight concrete in prEN 206 [4]

STRENGTH CLASS	$f_{ck,\ cylinder}$ [N/mm²]	$f_{ck,\ cube}$ [N/mm²]	DEFINITION
C12/15	12	15	Normal strength concrete
C16/20	16	20	
C20/25	20	25	
C25/30	25	30	
C30/37	30	37	
C35/45	35	45	
C40/50	40	50	
C45/55	45	55	
C50/60	50	60	
C55/67	55	67	
C60/75	60	75	High strength cconcrete
C70/85	70	85	
C80/95	80	95	
C90/105	90	105	
C100/115	100	115	

For the limitation of crack width, Eurocode 2 provides a classification of verification criteria which is presented in Table 7. The principle is that for a certain load level (infrequent, frequent, quasi-permanent) either the limit state of decompression or the limit state of crack width shall not be exceeded. The load levels in Table 7 have been derived from investigations and from experience. At the limit state of decompression, no tensile stresses in the concrete are allowed under the relevant combination of actions. It is relevant for prestressed members or for those subjected to significant compression forces.

For crack width control, the deemed to satisfy criteria given in Table 7 are based on the following design crack width w_k:

- For members with internal bonded prestressing tendons: w_k = 0.2 mm
- For members with reinforcing steel: w_k = 0.3 mm.

The categories A to E in Table 7 are chosen in relation to the environmental actions, the risk of deterioration and the design load level. The latter is mainly a function of the variable load, Q_k, which is defined in [1] as an upper value of a statistical distribution with an intended probability of 98 % of not being exceeded within a reference period of one year.

Values of Q_k can be found in Eurocode 1 (see Figure 1). The infrequent value of a variable action corresponds (approximately) to Q_k.

Table 7 Classification of verification criteria

CATEGORY	COMBINATION OF ACTIONS FOR THE VERIFICATION OF	
	Decompression	Crack width
A	infrequent	—
B	frequent	infrequent
C	quasi-permanent	frequent
D	—	frequent
E	—	quasi-permanent

The frequent value of a variable action, $\Psi_1 \cdot Q_k$, corresponds according to [1] to a value which is exceeded either 5 % of the reference time or 300 times per year. The highest value should be chosen. The corresponding combination coefficient, Ψ_1, will vary between 0.5 and 0.9, according to the variable action considered.

The quasi-permanent value of a variable action, $\Psi_2 \cdot Q_k$, which is commonly used for crack control in reinforced (i.e. not prestressed) concrete members, corresponds to the time average or to the value with a probability of being exceeded of 50 %. The corresponding value Ψ_2 will vary between 0.3 and 0.8.

Besides the above provisions which require numerical checks, a couple of deemed to satisfy rules are provided in Eurocode 2. They concern bar spacing, spacing of stirrups for the control of inclined cracks due to shear and/or torsion, a limitation of bond stresses and minimum reinforcement areas along the surface of concrete members. The latter are intended to resist self-equilibrating stresses and, thus, to ensure an adequate quality of the concrete at the surface of concrete members.

CURING

The durability of a concrete surface zone depends on several parameters, in particular on an adequate resistance against carbonation and a low-permeability. Both of these two parameters are a function of the degree of hydration which is a time-dependent process which depends mainly on the type of binder (cement) and on the ambient temperature during reaction. On the other hand, the strength development of the concrete is also a function of the hydration. An analogy between permeability *decrease* and strength *increase* exists. For this reason and in order to provide a practice-oriented engineering model, the concrete strength was used in [5] for the characterization of the degree of hydration after the curing period.

Therefore, according to [5] a concrete surface exposed to environmental conditions other than XO and XC1 in Table 2 shall be cured until the surface has achieved a certain percentage of the specified compressive strength. This percentage depends mainly on the moisture supply *after* curing, i.e. on the continuation of the hydration process after the curing period. In [16], corresponding values are given which vary between 60 % (very dry environment) and 10 % (humid environment). In [5], however, a constant percentage of 50 % is required for reasons of simplification.

This value may be considered to be achieved if the minimum curing periods recommended in Table 8 are applied. The Table applies to all environmental conditions in Table 4 other than XO and XC1 and distinguishes between different ambient temperatures t and the strength development of the concrete. This development is expressed by the ratio

$$r = f_{cm,2} / f_{cm,28} \tag{5}$$

where:

$f_{cm,2}$ is the mean value of compressive strength after 2 days

$f_{cm,28}$ denotes the mean value of compressive strength after 28 days determined from initial tests or based on known performance of concrete of comparable composition.

Table 8 Minimum curing period for EN 206 exposure classes other than XO and XC1

SURFACE CONCRETE TEMPERATURE (t), °C	MINIMUM CURING PERIOD, DAYS [1),2)] FOR A CONCRETE STRENGTH DEVELOPMENT $r = f_{cm,2}/f_{cm,28}$			
	$r \geq 0.50$ rapid	$r \geq 0.30$ medium	$r \geq 0.15$ slow	$r < 0.15$ very slow
$t \geq 25$	1.0	1.5	2.0	3.0
$25 > t \geq 15$	1.0	2.0	3.0	5
$15 > t \geq 10$	2.0	4.0	7	10
$10 > t \geq 5$[3)]	3.0	6	10	15

Note:
1. Plus any period of set exceeding 5 hours.
2. Linear interpolation between values in the rows is acceptable.
3. For temperatures below 5°C, the duration should be extended for a period equal to the time below 5°C.

For concrete surfaces to be exposed to exposure classes XO and XC1 in Table 2 only, the minimum curing period should be 0.5 day, provided the set does not exceed 5 hours and the surface temperature is equal to or above 5°C.

However, [5] does not exclude the application of other minimum curing periods provided that this can be justified by the conrete used and the curing method applied.

REFERENCES

1. EUROPEAN COMMITTEE FOR STANDARDIZATION (CEN). (1994) Eurocode 1: Basis of Design and Actions on Structures. Part 1: Basis of Design. CEN, Brussels. ENV 1991-1.

2. EUROPEAN COMMITTEE FOR STANDARDIZATION (CEN). (1991) Eurocode 2: Design of Concrete Structures. Part 1: General Rules and Rules for Buildings. European Prestandard. CEN, Brussels. ENV 1992-1-1.

3. EUROPEAN COMMITTEE FOR STANDARDIZATION (CEN). (1990) Concrete. Performance, Production, Placing and Compliance Criteria. European Prestandard. CEN, Brussels. ENV 206.

4. EUROPEAN COMMITTEE FOR STANDARDIZATION (CEN). (1997) Concrete. Performance, Production and Conformity. Draft. CEN, Brussels. prEN 206.

5. EUROPEAN COMMITTEE FOR STANDARDIZATION (CEN). (1998) Execution of Concrete Structures. Part 1: General Rules and Rules for Buildings. CEN, Brussels. Document CEN/TC104/SC2-N126.

6. COMITÉ EURO-INTERNATIONAL DU BÉTON (CEB). (1997) New Approach to Durability Design. An example for carbonation induced corrosion. CEB, Lausanne. CEB-Bulletin d'Information No. 238.

7. EUROPEAN COMMITTEE FOR STANDARDIZATION (CEN). (1994) Methods of testing cement - Part 2: Chemical analysis of cement. CEN, Brussels. EN 196-2.

8. DEUTSCHES INSTITUT FÜR NORMUNG E.V. (DIN). (1991) Assessment of water, soil and gases for their aggressiveness to concrete - Part 2: Collection and examination of water and soil samples. DIN, Berlin. DIN 4030-2.

9. INTERNATIONAL STANDARDS ORGANIZATION (ISO). (1984) Water quality; Determination of amonium - Part 1: Manual spectrometric method. ISO, Geneva. ISO 7150-1.

10. INTERNATIONAL STANDARDS ORGANIZATION (ISO). (1986) Water quality; Determination of amonium - Part 2: Automated spectrometric method. ISO, Geneva. ISO 7150-2.

11. INTERNATIONAL STANDARDS ORGANIZATION (ISO). (1986) Water quality; Determination of calcium and magnesium; Atomic absorption spectrometric method. ISO, Geneva. ISO 7980.

12. EUROPEAN COMMITTEE FOR STANDARDIZATION (CEN). (1992) Cement; Composition, specifications and conformity criteria - Part 1: Common cements. CEN, Brussels. ENV 1997-1.

13. HILSDORF, H.K. (1995) Concrete, published in Concrete Structures - Euro-Design Handbook, Ernst & Sohn, Berlin, pp. 1-103.

14. VISSERS, J.L.J. (1998) k-value for powder coal fly ash. Brussels, Internal paper of CEN/TC104/SC1.

15. DEUTSCHES INSTITUT FÜR NORMUNG E.V. (DIN). (1997) Concrete, reinforced and prestressed concrete structures - Part 1: Design. DIN, Berlin. prDIN 1045-1.

16. GRÜBL, P. (1996) European Concept on the Curing of Concrete; published in Concrete Precasting Plant and Technology, Vol. 62, Bauverlag GmbH, Wiesbaden, pp. 82-91.

CONCRETE RESEARCH IN PRACTICE – THE FOURTH DIMENSION

B J G van der Pot

HBG Group Technology

Netherlands

ABSTRACT. In the Netherlands the focus in concrete research is on "defined performance concrete". The aim is to put emphasis on developing knowledge to obtain defined performance, full control and no surprises rather than on ultra high strength and extreme performances. A brief synthesis is given of a research project called "Tolerant Concrete" which is focussed on developing an engineering concept which enhances the specification and control of performance properties in concrete structures.

Keywords: Defined performance, Particle packing, Self-levelling, Self-compacting, Rheology.

Ir Bart J G van der Pot is Director of HBG Group Technology in Rijswijk, the Netherlands. HBG is a large European construction group with the Netherlands, the UK and Germany as its home markets. Group Technology is the central department where technology initiatives are taken and facilitated.

INTRODUCTION

For a building contractor there are good reasons to be interested in concrete research. Most of the buildings and houses in the Netherlands are concrete based. Sometimes this is directly visible, but in most cases it is not because masonry facades or curtain walls of glass, steel, aluminium or natural stone hide the concrete columns, floors walls staircases etc. These houses and buildings are designed and built by building contractors who have to make some profit to enable them to stay in business. In order to do so, they have to be innovative, efficient and effective.

The relation between research at one side and performance, effectiveness, efficiency and ultimately profit at the other side is the combination I want to emphasise here. It is the link between concrete research and profit, or rather time, because in construction there is a strong relation between time and profit and an even stronger relation between lost time and lost profit. In most cases it are the building contractors who carry the cost for everything that goes wrong and requires extra time in construction.

In the title I have called this the fourth dimension, it is the dimension of time, money and usefulness, or in other words performance.

This dimension is reflected in the current way of thinking in concrete research in the Netherlands. In the last decade concrete research has, to a large extent, either been a sub-item of more integral research efforts into items such as underground construction, environmentally friendly construction etc., or it has been focussed on finding ever higher value applications in terms of higher strength or other properties. This research has obviously value, but it relates only to a low percentage of our business, it is directed to very special purposes and applications. For the majority of our daily work, all the houses, buildings and structures that we continuously produce, we need another focus; at least if we believe that there are still things to learn and to improve in these "normal" areas. In the Netherlands we have this belief and we have called this focus "Beton op maat", probably best translated by "defined performance concrete", as a contrast to the "high performance" approach.

Last year CUR, which is the Dutch organisation which co-ordinates and programs the concrete research activities in the Netherlands, has installed a programme advisory committee on concrete research. This committee has come to a programme approach with a focus on "defined performance concrete".

What we want is probably less spectacular, but more useful; not ultra high strength or extreme properties, but full control and defined performance. We want to be able to properly plan our work and have no surprises whatsoever; we want to provide our customers with a guaranteed product in full conformance with their specifications and expectations as well as a lot of other things which are not spectacular but can improve the speed, the control and the quality of our products and working circumstances.

Ideally concrete, as a material, has to be able to comply with an endless amount of requirements which vary from application to application and from project to project. As a result of all the research and development over the past decades, concrete has become a very complicated and knowledge intensive material. It can be given an endless variety of properties by varying its composition, additions or treatment.

Knowledge has become a vital ingredient for getting the most out of the material, for being commercially successful and for enabling creative thinking. The average user of the material, either in the design, specification, production, procurement or execution phase, only knows a fraction and only has access to a fraction of the available knowledge which could help him to prevent problems or to bring him to innovative solutions or improvements. As part of our "defined performance" approach we also have identified knowledge and knowledge technology as key attention areas.

Hereafter I give a short briefing of a Brite Euram research project named Tolerant Concrete in which we have participated and which has this focus on defined performance, full control and no surprises.

TOLERANT CONCRETE

Scope

During the last decade a large number of civil engineering structures were built using concretes which were designed for excellent properties with regard to production (placing and compaction), structural behaviour and durability. In several cases however these objectives have not been realised. Part of the reasons can be traced back to the consolidation phase of the concrete. They showed up in e.g. segregation in macro and microstructure of the concrete, intrinsic cracking and bad air void structure.

This resulted in unforeseen repair and increased maintenance costs and reduced service life as well as extra costs during construction resulting from rejections, corrective actions and delays.

These deficiencies are basically caused by the high sensitivity of the concretes in the consolidation phase during placing and compaction and in the early phase where the concretes, due to low strain capacity are sensitive to thermally, chemically and mechanically induced stresses and displacements. One of the observations was that in laboratory circumstances, with small specimens, a level of control over the properties could be achieved which in real circumstances could not be met at all times

The low tolerance of concrete for all kind of variations and occurrences has resulted in costly problems and repairs.

To solve this problem the BRITE EURAM proposal 'Tolerant Concrete designed for high performance during production and structural service life' was granted by the EC. The Tolerant Concrete research project was executed from 1994 till 1998.

The objective of the project was to develop an engineering concept to enable the designer of concrete to specify concretes with clearly defined performances being:

- tolerant to minor errors during production and curing and to the type of mineral aggregates; the choice of aggregate is largely dependent of local availability because large amounts are needed and transport costs may therefore be a substantial part of the material costs

- easy to compact with low or no energy consumption; related types of concrete are the so called flowing concretes, self levelling and self compacting.
- low sensitivity to intrinsic shrinkage
- improving the working condition of labourers
- reducing noise production during compaction

Gains were expected to be:

- higher production reliability
- improved durability due to the homogeneous and dense structure
- increase of the production rate
- reduction of the impact on labourers
- general improvement in environmental conditions and consumption of resources.

It was aimed to develop a concrete which requires no or highly reduced energy for compaction, which is structurally stable during transportation, pumping, placing and consolidation, which does not become inhomogeneous in the micro structure during setting and which lends itself to a predictable and gentle curing. In short: risks in execution should be avoided and there should be full control on the properties during all phases.

This new type of concrete had to be based on actual theories concerning the effects of combining minerals and silica fume spheres in the cementitious binders. To achieve this objective it is necessary to establish an understanding of the interaction between these micro particles and to develop dispersing agents, which at the same time work on all constituents.

Project

The research project was executed by a consortium consisting of 8 partners:

- Dansk Beton Technik A/S [DK]
- Geological Survey of Denmark and Greenland[DK]
- Demex A/S [DK]
- English China Clay Int. Ltd [UK]
- Hollandsche Beton Groep NV (HBG) [Nl]
- Intron BV [Nl]
- Eggerding & Co Groep [Nl]
- Perstorp Chemitec AB [SE]

The project started in 1994 and finished in 1998. The 12 project tasks can be clustered in 3 groups:

- identification of the research topics
  - definition of the ‘high performance concept’
  - identification of related research and patents
  - identification of commercially available clays in Europe and description of their characteristics
  - identification of surface active agents which may act as dispersers for micro particles in multi powder combinations at high pH-levels

- development of techniques and understanding
  - optimisation of binders
  - ultra fine grinding technologies
  - verification of the basic theories of the research
  - development of 6 different types of concrete based on the verified principles
  - development of low compaction techniques

- verification
  - verification of rheological and functional properties of the new type of concrete through trial castings
  - assessment of the impact on the entire production process on working conditions and external environment compared to traditional production
  - documentation of properties in full scale structures

Principles

The aimed performance is strongly related to an optimal grain packing on all length scales. This optimum must be obtained in the production process. Important are therefore:

- the quality and proportions of the constituents
- storage, mixing and transportation of the fresh mix
- moulding, placing and compaction

Regarding the constituents three principles may be considered:

- Extending the principle of underwater concrete (wash out concrete or non-dispersive concrete) which is self-compacting and non-segregating. The technique is based on addition of chemical admixtures to improve the viscosity and thereby preventing dissolution of particles in the surrounding water. The result however is a sticky mix, which may result in inhomogeneous concrete, especially near rebars etc. Also the high cement content may result in unfavourable heat development and high material costs and furthermore the use of large amounts of superplasticiser may lead to fast loss of workability, retarded setting time, low early strength and further increase of material costs.

- The second concept, which is mainly due to Japanese research, is based on optimisation of the sieve curves of traditional materials. The relative volume of aggregates is reduced to prevent agglomeration of aggregates when the concrete passes obstacles. The powder to water ratio was adjusted to approximately 1 to produce an adequate viscosity to avoid segregation and to obtain self-levelling properties. The powder implies large amounts of mineral particles like slag, fly-ash and mineral fillers of e.g. limestone and quarts. This concept is risky. Noted are a high sensitivity to overdose of superplasticiser, variations in content of water adhering to the relative fine mineral aggregates and a high internal material disorder.

- The concept followed in this Brite Euram project is to add mineral constituents to realise the desired properties without having the above mentioned disadvantages. Theories are hardly available so the research bears an empirical character.

All three concepts are based on optimisation of the rheological behaviour of the species in order to avoid separation and to reduce compaction effort and finishing works. The use of super plasticisers is inevitable.

The usual defects in concrete are of the following nature:

- bonding defects at coarse aggregate and reinforcement, mainly evolved as highly porous zones of varying thickness
- uneven homogeneity of the cement paste, i.e. varied porosity
- cracks, both macroscopic in exposed surfaces and microscopic in the volume
- uneven distributions of constituents
- poor quality of the air void system (low specific surface)
- local lack of hydration of cement grains

For these defects two main causes may be noted:

- instability at micro – and macro level in the fresh concrete
- inappropriate handling, primarily excessive vibration

Particle Packing

It has been recognised for many years that, besides grading of aggregates, the particle size distribution on micro level is important for the properties of the slurry and the concrete. Therefore calculations have been made to get insight in the optimal composition, which is defined as the one with the highest packing density. As the numerical effort is considerable the calculations were restricted to 2 dimensions. Two models have been developed. In a kinematic model only geometrical compatibility of circular elements, which are the 2 dimensional equivalent of spheres, is demanded. A more advanced dynamic model was subsequently built to take into account forces between particles, like gravity, attraction- and elastic forces, and constitutive properties, like elasticity and viscosity. These models were made available for computerised use in the software program FEMMASSE. The steps in the calculations are:

- generation of a particle structure (suspension, slurry or solid) according to the applied material laws
- link mechanisms to the generated structure
- optimisation with regard to the criterion of optimised density as a consequence of particle packing.

It appeared to be possible to optimise the composition by means of the developed calculation scheme and the founding assumptions. This mix was subsequently made in the laboratory and appeared to behave well.

Dispersion

One of the conditions to arrive at an optimal packing of the particles is an optimal dispersion of the fines during the fluid stage of the mix. A dispersion agent may furthermore act like a plasticiser for the cementitious part of the binder. Demands posed on the mix from practice may be formulated as:

- workability of the mix for more than 3 hours
- pumpability for more than 2 hours
- initial setting time less than 6 hours.

The theoretical knowledge is rather well known as well as the action of candidate dispersing agents:

- disperse materials in water
- stabilise the dispersion

Due to the stabilisation workability is increased and setting is retarded, so demands on these aspects may be contradictory. These agents usually are of inorganic nature and are usually based on for example naphthalene or melamine. They lower the viscosity and thereby allow reduction of the water content and thus result in strength increase of the concrete. A major obstacle for more complex mixes becomes that the dispersion affinity of a type of disperser differs for different powders.

Good flow properties and high levelling and filling capacity are key properties for high performance concrete.

Grinding

As grinding of slag, cement and minerals is of great importance for the success of the promoted addition, methods, fineness and accuracy of grinding processes have been investigated. A typical classification according to resulting size is:

- crushing $100\% > 200\ \mu m$
- grinding $100\% < 200\ \mu m$
- micronising $100\% > 7\ \mu m$
- sub-micronising $100\% < 7\ \mu m$

Another classification is to grinding principles:

- crushing
- impact
- attrition
- scraping / shearing

Furthermore a large number of grinding mills have been discussed and tried out and a large number of selection criteria have been considered. The effect of micronising the additives is to obtain a better build-up of the paste i.e. optimal filling of the pores. Since these additives are not hydraulically active, wet grinding is the most suitable technology. To obtain a fineness of 1 μm an eccentric vibration mill or a pearl mill is necessary. Otherwise the higher capacity dry ball mill can be used. For micronising hydraulic components dry ball milling process is recommended.

Next to size also shape may influence concrete strength and rheological behaviour. Experimental determination of these effects is obliged for optimisation.

Selection of Mixes

After a first selection of mixes, candidate mixes have been identified by using laboratory tests. The chosen mixes have been evaluated at this stage of the research by semi full-scale tests.

- On beforehand the use of 3 types of cement (CEM I 42.5 from Aalborg, CEM I 32.5 R from ENCI and CEM III/B 42.5 from CEMIJ), 4 ranges in water-cement ratio (0.35 – 0.55) and use of 2 types of superplasticisers (melamine and naphthalene based) was foreseen. From the possible combinations a selection was made. The aim of the first part of the research was to select a small number of candidate operational mixes with locally available mineral aggregates

- The optimal mixing procedure was found to be dry mixing of all powder components and aggregates, addition of water and finally step-wise addition of superplasticiser. A practical viscosity-measurement apparatus, the concrete tester, was found to be indispensable for the determination of the amount of superplasticiser to be added.

A second series of tests has been carried out in Denmark with locally available mineral aggregates. The objectives were to test mix designs and mixing procedures with regard to the demands for tolerant mixes:

- workable under difficult conditions
- transportable under vibration without loss of homogeneity
- low compaction effort needed

The mixes were prepared with great care in order to eliminate differences arising from the preparation procedure. The resulting mixes and concretes have been evaluated with respect to:

- workability (slump and flow test)
- air content
- density
- cohesiveness, homogeneity, bleeding (subjective by inspection by bare eye and by microscopic evaluation of epoxy impregnated plane sections)
- compressive strength and Young's modulus
- frost resistance
- chloride diffusion and permeability
- petrographic and air void analysis

It was concluded that all three mixes are promising for developing an operational tolerant concrete.

Full Scale Trials

The aim of the full-scale tests was to verify and to demonstrate the practical applicability of the mixes, which acted well in the laboratory tests with regard to:

- less or strongly reduced vibration for compaction
- avoidance of aggregate pockets
- smooth surfaces
- homogeneous product

As a realistic but hard to fabricate building, a L-shaped mould with a volume of 7 m^3, a height of 3.3 m and the inclusion of a casting joint was chosen. The test was executed on side of a mixing plant with two batch capacities of 2.5 m^3 and a production capacity of 160 m^3 / h. Local gravel aggregate from the North Sea and the blast furnace slag and Portland fly-ash cements available at the plant were used.

The full-scale trials have shown that self-compacting concrete based on the developed theory can be made on a practical scale. Both blast furnace slag and Portland cement may be used, as well as aggregate of different environmental origin. Critical aspects are:

- the batch size should be at least 1 m^3 as smaller batch resulted in a relatively large amount of the sticky binder on the batch walls
- the type of aggregates has a large impact on the mix properties
- provisions for avoiding air entrapment near the mould; entrapped air may probably be a reason to apply reduced (mould) vibration
- the pumpability.

Actual Test

Having concluded from the full-scale test that the concept of tolerant concrete is practical and workable –at least regarding the flow and compaction properties- an actual building site was sought. In Uelzen a local river underpasses a canal and for several reasons the siphon had to be reinforced. So a new concrete wall had to be built against the existing one. This new wall was to be built in segments, which were to be casted one at a time. It was foreseen to make a number of them in tolerant concrete.

Before applying the concrete in the segments of the siphon, two test segments have been cast just outside the siphon. These segments had been reinforced heavily. Thus the full industrial process has been simulated, including the mixing, the transport and the use of raw materials. The casting was done by pumping over a distance of 50 m. The mixes used in both casting were almost identical.

After demoulding it appeared that the findings of the full scale test were confirmed. The bulk material was (very) satisfying, but the surface was of less quality. In the casting of the first mould the pump pressure was low, in order to apply the concrete very gentle. This in contrast with the full scale test. In the second trial the pump pressure was normal.

It is observed that the smoothness of the surface was very good in case the Zemdrain was used. At the surfaces were Zemdrain is not applied, smooth areas and areas with air and water inclusions could be recognised. A correlation with (vertical) movements of the casting pipe has been recognised.

Temperature measurements all over the mould have shown a maximum increase of the temperature during hardening of 20 °C.

From the tests it was concluded that the tolerant concrete principle works and is able to yield good mixes, although the application technique may require some development. The mix itself has very good flow properties and is self-compacting. Slight mould vibration could be helpful to obtain smooth surfaces

CONCLUSIONS

The results of the Tolerant Concrete project supports the "defined performance" approach with respect to concrete research.

The overall conclusions of the Tolerant Concrete research project are:

1. the developed product is tolerant with regard to handling, though some aspects may not be disregarded (minimal batch size, form work provisions) but not to changes and variations in composition

2. no compaction energy is needed, except in case light mould vibration is chosen as the form work provision for obtaining smooth surfaces

3. mechanical properties, including shrinkage, can be accurately realised

4. as a consequence of the avoidance of vibration labour conditions and noise production during execution are seriously improved

5. as a consequence of the well controllable properties durability of structures can be improved by using the developed mixes

REFERENCES

BRITE EURAM. T BRE2-CT94-0972, 1993. Tolerant Concrete designed for High Performance during production and structural service.

WHOLE LIFE DESIGN FOR DURABILITY AND SUSTAINABILITY. WHERE ARE WE GOING, AND HOW DO WE GET THERE?

G Somerville

British Cement Association

United Kingdom

ABSTRACT. It is argued that durability design is an integral part of a move towards greater sustainability. The definition of performance requirements is an essential first step, while relating technical performance more to the future operational requirements for the structure. There is also a need to radically review design methods - a cultural change, requiring less dependency on numerical analysis and more on providing satisfactory whole life performance at minimum total cost. This paper reviews the options for change, summarizes the major difficulties, and presents a strategy for future development - with the emphasis on a design-led holistic approach.

Keywords: Sustainability, Durability, Integrated design, Performance requirements.

Professor George Somerville is Director of Engineering at the British Cement Association, and a Visiting Professor, both at Imperial College, London, and at Kingston University.

INTRODUCTION

In 1986, the author [1] made a distinction between the production and placing of durable concrete, and the design and construction of structures which would be durable. The essence of the argument was that we tried to solve our durability problems via a prescriptive materials approach, and yet standards of design, detailing and construction were at least as important in practice.

Except for checks against Code requirements for concrete cover and grade, durability was not an integral part of the design process, which was perceived as the provision of adequate strength, stiffness, stability and serviceability - largely a numerate process, following well-trodden paths laid down in Codes of increasing complexity.

In the interim, that scenario has changed little - at least for routine everyday practice. However, there have been significant trends and developments, which, in time, could and should radically change the whole basis of design - both in terms of performance requirements and in our approach to satisfying these requirements. These developments include the following:-

- A growing demand, on the part of owners, for structures which better meet their operational needs, while recognising the importance of life cycle costing and the reality of planned maintenance, management and replacement;

- significant improvements in our ability to specify durable concretes for specific aggressive actions;

- the introduction of a wide range of relatively new protective measures for corrosion resistance (coatings, sealers, layers, inhibitors etc.);

- the emergence of a better perspective of the relative importance of the different deterioration mechanisms, and of predictive models to quantify their effects - mainly via more definitive state-of-the-art reports (eg.[2-5]);

- major attempts to develop numerate design concepts for durability, which differ only in how they deal with reliability, safety and margins (eg. [6-11]). Here, the approaches have a strong emphasis on design, while recognising to varying degrees the importance of a holistic approach, which also takes on board material, construction (quality) and maintenance issues;

- a relatively recent recognition of the need for greater sustainability, and the significance of durability design, in meeting that need.

There is little doubt that the groundswell created by these developments will accelerate a move towards rational durability design. What is less certain is the form that that should take.

The purpose of this paper is to examine the options, while identifying the areas of difficulty in taking these forward, before suggesting a strategy for future development.

THE BIG PICTURE - SUSTAINABILITY, DURABILITY AND PERFORMANCE REQUIREMENTS

In a broad, even global sense, sustainability is sometimes simply defined as:-

"Development that meets the need of the present, without compromising the ability of future generations to meet their own needs"

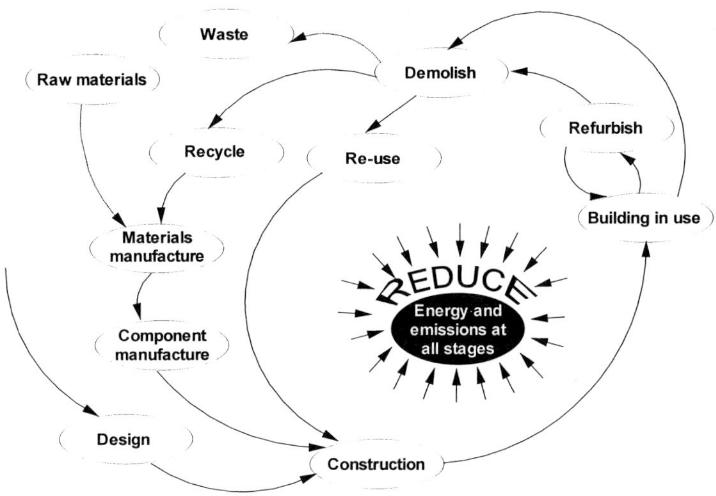

Figure 1 Material flows; cradle to the grave

This definition embraces climatic, environmental, social and economic issues. For construction, these translate into factors which include:- energy conservation, waste reduction, improved quality of life and environment, more efficient use of scarce or non-renewable resources, recycling, better operational performance, etc.

A simplistic, but commonly used, illustration of sustainability in construction is the cradle to the grave material flow shown in Figure 1. In particular, this emphasises the current focal points for sustainability - energy and emissions, with some interest in recycling; better use of waste materials is also an integral feature.

However, in the present context of durability and design life, a more relevant sustainability diagram is that shown in Figure 2, which plots 'sustainability' against 'quality', where quality is representative of a desired achievement in design, compared with required performance in service.

In simple terms, this suggests that we need to move from curve 1 (representative of present practice, where sustainability is not yet a driver) towards curve 2, thus creating a zone of sustainable construction with a definite plateau.

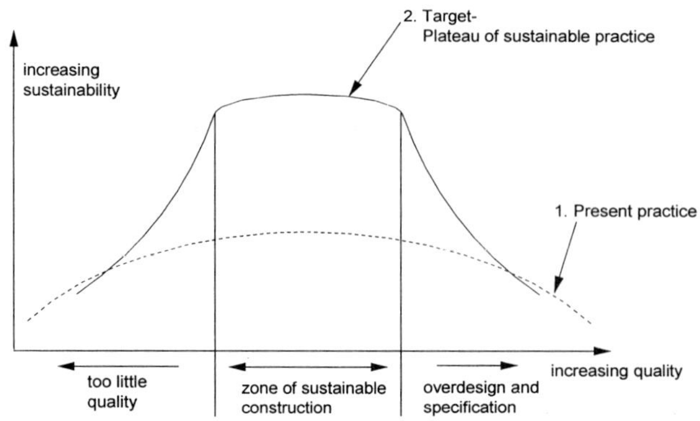

Figure 2 The plateau of sustainable practice

If, on the one hand, we put too little quality into our structure, and it fails to meet the owner's in-service performance requirements, then this constitutes 'failure', in sustainability terms. On the other hand, it is equally bad to over-design or over-specify, since this wastes scarce resources. The only solution to this dilemma is to get as close as we can in design, to an exact fit with clearly defined performance requirements.

Figure 1 presents an outline of the sustainable construction arena. Figure 2 gives the elements of a strategy, which can lead to a broad definition of performance requirements. In fact, that given in the CEB - FIP Model Code [12] fits within the sustainability concept; this is as follows:-

> *"Concrete structures shall be designed, constructed, and operated in such a way that, under the expected environmental influences, they maintain their safety, serviceability and appearance during an explicit or implicit period of time, without requiring unforeseen high costs for maintenance or repair"*

This represents a move towards a performance-based design approach, consistent with greater sustainability. Essential to this is the establishment of limiting performance criteria; this has to be at 2 levels:-

1. A requirement, which deals with the in-service time factor - possibly a target, somehow expressed in terms of 'life';

2. A requirement which directly relates to a minimum acceptable technical performance, over that life.

The 'pure' approach to 1.[6][7] is to use a reference period of, say, 50 years for buildings and relate it to a reliability index of 3.8 for the ultimate limit state (as given in ENV 1991-1) and of 1.5 for serviceability. This sets a mean service life, over which the normal imposed loads and load combinations are to be carried. This can lead to service lives which are well beyond our experience. It also begs the question of what constitutes minimum technical performance, due to effects caused by deterioration mechanisms, and how these might affect normal action effects such as bending and shear.

In considering how to move forward here, a broader view is necessary. Most cities, in most countries, have buildings dating from the last century; few will be exactly the same, due to changes in use, upgrading, new cladding, even some replacements in whole or in part. Obsolescence, and change in use, is therefore an issue.

To some extent, the situation is similar, for civil engineering structures, especially for highway structures on principal routes.

In addition, based on experience with modern infrastructure, owners are now more conscious of the need for regular inspection and good maintenance. Further, if structural changes have to be made for whatever reason, they wish the disruption to the operation of the structure to be kept to a minimum.

In developing this type of management and maintenance strategy, it is important to remember that a structure is not a single unit; it is made up of numerous elements, services, furniture and fittings - all of which have different lives.

Table 1 Basics of a performance profile plan for building

SYSTEM	CRITIC-ALITY	TARGET LIFE BEFORE REPLACEMENT, YEARS						CAPITAL COSTS*	COST IN USE TARGET: X YEARS AT Y PRICE**	MAINT-ENANCER EQUIR-EMENTS
		>5	5-10	10-20	20-40	40-100	>100			
Foundations	A	••••••	••••••	•••••••	•••••••	••••••	•••	—	—	Clear specif-ication for each system
Structure	A	••••••	••••••	••••••	••••••	•••		—	—	
External Cladding	B	••••••	••••••	••••••	••••••			—	—	
Glazing	B	••••••	••••••	••••••	•••			—	—	
Partitions	B	••••••	••••••	••••••	•••			—	—	
Heating & Ventilation	B	••••••	••••••	••••				—	—	
Water, Public Health Service	A/B	••••••	••••••	••••••	••••••			—	—	
Electrical	A	••••••	••••••	••••••	••••••			—	—	
Decorating	C	••••••	••••••					—	—	

* e.g., 100 units
** e.g., 1500 units over x years

The trend here is to plan for this in a systematic and economic way, using performance plans, shown simplistically in Table 1 for a building. Doing this properly involves whole life costing methods, with the objective of minimising total cost and maximising the operational use of the structure.

This type of thinking (obsolescence, systematic performance plans) promotes the idea that design life is not a single absolute value, but will vary, depending on the type of structure, on its use and on the maintenance/replacement plan (Table 1). Actual values will be established, in individual cases, at the conceptual design stage, between the client and the designer. The designer must then have available to him:-

a) A simple classification system for distinguishing between the life time requirements for different elements in the structure as a whole.

b) A design procedure which ensures that the performance requirements are met for each classification.

One possible classificaiton is:-

A. *Life-long* The element is so important that failure would cause cessation of function, and/or major disruption or cost, during remedial work. Highly critical.

B. *Repairable* Efficiency of operation would be reduced, but replacement/remedial work could be done economically. Medium criticality.

C. *Replaceable* Routine maintenance or replacement could be done at minimum cost and little inconvenience. Low criticality.

A British Standard exists [13], based on this principle, which simultaneously encourages the development of materials and components on a targeted service life basis; similar approaches are under development at a European level via CIB W60 and RILEM PSL140.

The key point being made here is that required life and performance should relate to the owner's operational plans and maintenance plans. It is often assumed that 'better' durability means longer life. This is not so. The sustainability argument implicit in Figure 2 makes it clear that the design objective is to match, as best as we can, the performance profile in Table 1.

A QUANTITATIVE APPROACH - MAJOR HURDLES TO BE OVERCOME

General

The first major hurdle - the definition of performance-based criteria - has been discussed in the previous section; real progress is being made here. What other difficulties exist? To appreciate this, it is necessary to resort to feedback from in-service performance. Figure 3, from reference 10, presents typical data.

Easily the most dominant mechanism is corrosion, from various causes. However, there are variations between the different types of structures. External chlorides dominate for both bridges and marine structures, and are the most significant contributor for car parks; for buildings, corrosion is mainly initiated by carbonation of the concrete.

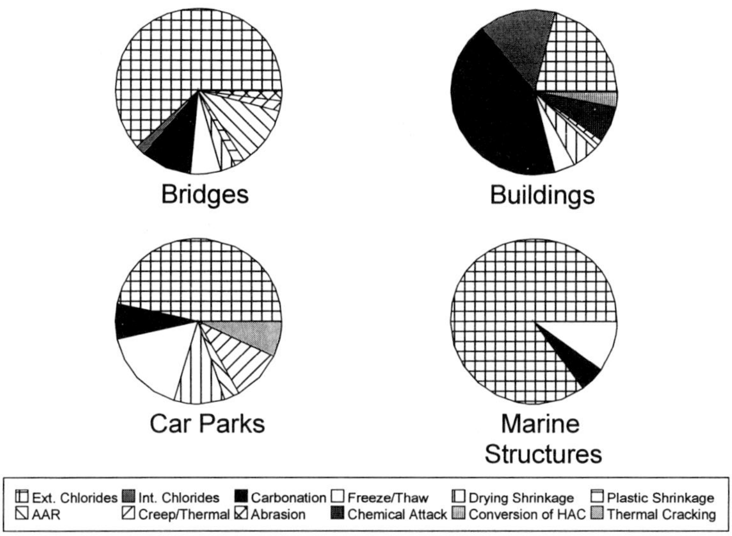

Figure 3 Reported Deterioration Mechanisms for Bridges, Buildings, Car Parks and Marine Structures

Table 2 Factors contributing to the failure of all structures reviewed in reference [10]

FACTORS CONTRIBUTING TO DETERIORATION	NO. OF CASES	NO. OF CASES %
Low Cover	47	11.6
Environment	156	38.5
Poor Quality Concrete	64	15.8
Poor Design Detailing	29	7.2
Poor Workmanship	17	4.2
Wrong specification	6	1.5
Failure of joint/waterproofing	31	7.7
Inadequate conceptual design	2	0.5
Wrong material selection	53	13
Total Number. of Cases	405	100

Of particular interest, in reference 10, is the analysis of the factors contributing to the deterioration. This is shown in Table 2. 'Environment' (natural or man-made) is the biggest single factor. However, if Low Cover, Poor Concrete Quality, Poor Design Detailing, and Poor Workmanship are added together then, collectively, they are of the same magnitude as for Environment. This feedback strongly suggests that, if durability design is to be significantly improved, two essential elements must be dealt with:-

1. Identification and quantification of critical aggressive actions.

2. Improving, or otherwise dealing with, quality of construction.

Quantifying Aggressive Actions

It is possible to do this for individual aggressive actions which directly attack the concrete. Table 3 shows the general approach which has been developed in Codes and Standards. This is a specification approach, which involves defining ranges of intensity for each action, and providing a corresponding prescriptive solution.

The major difficulty arises in dealing with corrosion. It is not difficult to identify the major causes of corrosion - chlorides and carbonation - but the effects of these, in penetrating the protective concrete cover, are much influenced by general and local climatic conditions. This involves an almost infinite combination of moisture, wind and temperature, creating a wide range of transport mechanisms of varying intensity.

As part of that, there is a strong interaction between the local climatic conditions and the outer fabric of the structure itself. Any planes of weakness (joints, cracks, etc) will quickly be found out - and therefore there is a link between durability and structural design and detailing.

Table 3 Types of aggressive action for which material specifications have been developed

AGGRESSIVE ACTION	GENERAL APPROACH	COMMENTS
Sulfate attack	Quantify the action	Specific material and mix proportions are recommended in most codes for defined ranges of sulfate concentration.
Alkali-silica reaction	Define ranges of intensity for it	The basic reaction and its possible effects are now well understood. Recommendations to minimize the risk of damage are published.
Freezing and thawing	Produce a specification for each range	Dealt with by choice of materials, mix proportions, and concrete grade. Air entrainment for lower grades. Detail to minimise exposure to moisture.
Abrasion		Specifications to cover aggregate properties, concrete grade and mix proportions, compaction and curing, methods of finishing, etc.

Table 4 Factors, and interactive effects, which influence 'loads' due to external chlorides

C2 Environmental loads due to chlorides

Actions
1) Chloride concentration. Type of cation
2) Humidity - Temperature
3) Oxygen concentration at rebar surface
4) Carbonation [hydroxide content]

Influencing concrete properties
a) Permeability [Diffusion coefficient]
b) Binding capacity [C_3A content]
c) Concrete humidity content
c) Alkalinity [Cl^-/OH^- threshold]

Climate classification

Micro Actions

A.	Wind direction
B.	Wash out from rain
C.	Condensation:- Sun orientation, Temperature, Closed spaces
D.	Leaks - Joints
E.	Irregular distribution of chlorides

Macro environment

I.	Marine	- Air - sheltered, exposed
		- Submerged
		- Tidal
		- Splash
II.	Deicing salts	- low cycling [Kg NaCl/year], frequent cycling
III.	Industrial	

How can we deal with this? Any attempt at an environmental classification, sufficiently simple for general use in Codes, is unlikely to represent real conditions. The real world, in terms of micro-climate and interactive effects, is typified by Table 4. Translating that into meaningful general Code guidance is difficult, as has been demonstrated by the CEB, in their most recent Bulletin on durability design [6].

Following on from that, we have the difficulty of dealing with the interactive effects between the aggressive actions and the outer fabric of the structure. This is implicit in Table 4, and shown graphically in Figure 4.

It is difficult to see a unique universal solution to this problem, and it may have to be addressed in different ways, as we shall see later. Nevertheless, one general principle can be stated here:- the need to consider water (moisture) alongside wind and temperature, in general design procedures - and especially for the detailing of facades in buildings and the outer exposed surfaces in other structures.

All structural engineers now consider wind and temperature in their designs (although neither is fully treated in Codes).

They would be concerned with safety and with stress limits, and, to some extent, with movement. Movement can lead to deformations and cracking, thus creating openings for wind-driven moisture. Removing, or keeping out water is fundamental to improved durability. It is a transport mechanism for aggressive actions, and the environmental conditions in Figure 4 can be translated into a range of penetration mechanisms, which involve pressure, capillary action, surface tension or gravity.

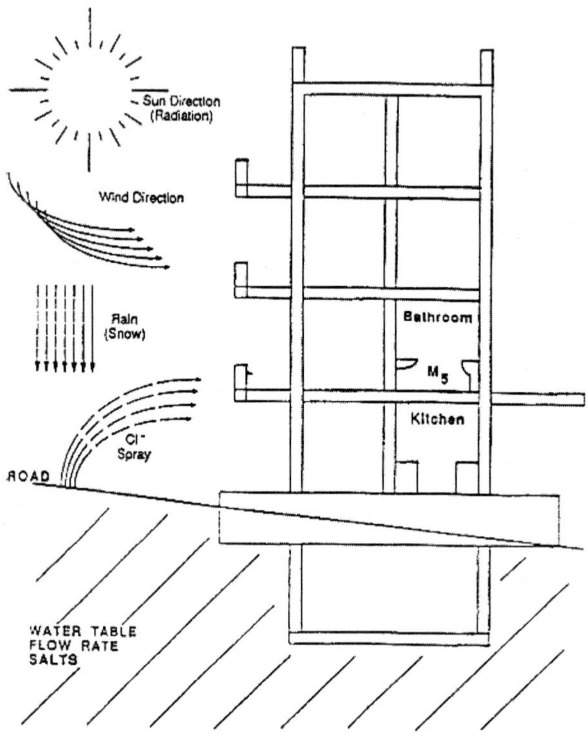

Figure 4 Possible influences on the outer micro-climate for a building

What is being described here is a fundamental difficulty in truly representing the conditions in Table 4 in Code terms, followed by finding detailing solutions to the outer fabric or surfaces of structures which either eliminate the effects of the environmental loads in Figure 4, or control them in a way which can be reliably taken into account in design. A durability design approach has to take these issues on board.

The Influence of Construction Quality

Figure 3 and Table 2 clearly illustrate the importance of construction quality and variability, particularly in providing resistance to corrosion. Space does not permit a thorough review of this subject, although it has been covered in a recent report [14].

In this context, variability relates both to the achievement of concrete cover, and to the quality of concrete itself. It should be recognised that there are only two basic ways of addressing the issue, in developing a quantitative generic solution. These are summarised in Table 5. Option 1 involves improvement, via the development of a quality package, within a systems approach, which raises standards to an acceptable (and repeatable) level. This should be an industry objective in any case. However it should be stressed that improvements must take place in design and detailing, as well as in site practice. Some design solutions are extremely difficult to build; others are deficient in providing adequate planned resistance to aggressive actions [14].

Table 5 Options for dealing with construction quality in durability design

OPTION	DESCRIPTION
1.	Raise standards - encourage better design and detailing (buildability) - improve communications from designer to site operative - education and training - bench marking best practice - method statements and certification
2.	Assume current variability - develop better defence strategies - introduce multi-layer protection systems - make more use of a ‘design-out’ approach

THE DESIGN APPROACH IN OUTLINE

Traditional structural design can be conveniently considered in two phases:-

1. Initial conceptual design, associated with preliminary sizing of elements and costing.

2. Final detailed design, essentially a numerical process, involving structural analysis, section design, and satisfying prescribed limit states.

Conscious design for durability has somehow to be fitted into this scheme of things. How can that be done? To begin with, it is probably necessary to subdivide phase 2, as follows:-

(2a) Detailing, either in terms of reinforcement arrangements, or the final shaping of cross-sections (for both buildability and durability), or integrating the basic structure into the artefact as a whole (e.g. cladding, services, finishes, etc., for buildings).

(2b) The numerate processes associated with limit states.

The prime reason for doing this is that phase (2a) is likely to be much more significant in durability terms, and it is not yet clear how valid detailed calculations might be, in satisfying defined durability limit states for phase (2b). In addition, conceptual design will be much more important for durability.

PHASE 1 - CONCEPTUAL DESIGN

The basic issues requiring consideration at this stage are listed in Table 6. It is intended to focus on section D in particular, in developing a defence strategy, while noting that section E will also influence this, both in terms of costs and in making judgements of possible levels of construction quality and of the known performance of the materials and components to be used.

Table 6 Conceptual design - issues and approach

A	Function and type of structure	(i)	Client's basic needs
		(ii)	Overall requirements (dimensions, etc.)
B	Performance requirements	(i)	Safety and serviceability criteria; minimum acceptable technical performance
		(ii)	Importance of continuity of function
		(iii)	Nominal design life (structure and/or elements)
		(iv)	Management and maintenance strategy
		(v)	Likelihood of future change of use, or upgrading
C	Loads	(i)	Dead and imposed loading
		(ii)	Wind, water and temperature effects
		(iii)	Aggressive actions
D	Overall design approach	(i)	Basic defence strategy
		(ii)	'Design out' versus 'provide resistance'
		(iii)	Articulation system: movement & joints
E	Preliminary evaluation of alternatives	(i)	Costs, including whole life costs
		(ii)	Construction method and quality issues
		(iii)	Known performance of materials and components

The essence of section D is to formulate a basic defence strategy. If we assume that we can deal with most aggressive actions that can directly affect the concrete, via a prescriptive approach, then the prime concern is corrosion, possibly affected by synergetic effects from other actions. The options for a defence strategy are then those shown in the flow diagram in Figure 5. None of these options are completely separate; the boxed feature for each of the three primary routes merely indicate the main component.

How might this work in practice, and which approach might be preferred in any given situation? To some extent, this will depend on the confidence that the designer has in:

1. His knowledge of possible aggressive actions, and the general and local environmental conditions.

2. The reliability of the various resistance options at his disposal.

An outline of some possible scenarios may illustrate the type of thinking required.

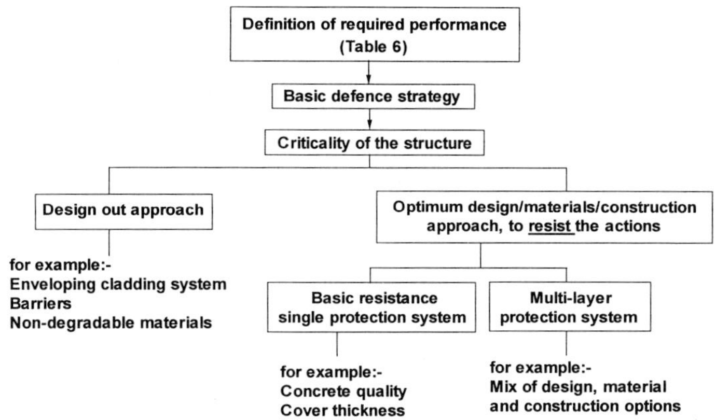

Figure 5 Options for a defence strategy, in durability design

Buildings in General

A design out approach may be attractive here, since it is comparatively easy to provide an enveloping cladding system, and the structural concrete would then only require a basic resistance appropriate to indoor conditions. Alternatively, special attention could be given at the phase 2a design stage to the detailing of the facade, as part of a multi-layer design approach. In this instance, movement and joints are important considerations.

Civil Engineering Structures in General

The design out approach is less practical here as a single protective option, although both barriers and non-degradable materials can be part of a multi-layer protective system especially for critical or vulnerable zones. The most common approach will therefore be a resistance option of some kind, whose nature may depend on the type of structure, the scale of the project, and what is known about the nature of the environment (which is commonly severe). Some examples are given below.

Maritime Structures

The performance of concrete in maritime situations has been well researched, and specialist guidance is available, in terms of design and specifications for the critical tidal and splash zones. Based on this, general approaches can be developed, most probably via the multi-layer protection route in Figure 5.

Major Structures

If the scale (and cost) of structures is very large, then special development work can normally be undertaken to produce solutions unique to the conditions. Two recent examples illustrate this. These are the Oresund crossing between Sweden and Denmark [15], and the Confederation Bridge in Canada [16].

Other Structures

For all other structures, but especially civil engineering structures in severe environments, where the local micro-climate may be difficult to define with sufficient precision to permit in-depth modelling and analysis, a multi-layer protection approach will usually be necessary.

PHASE 2A - DESIGN DETAILING

There is little to add about this phase, at this time. It is essentially concerned with the issues raised in Table 4 and Figure 4, associated with the need for designers to consider water alongside wind and temperature in evolving details, which create effective barriers to transport mechanisms. It does not involve intense numerical calculations or modelling. It does involve building up (and using) a portfolio of good practical details, based on feedback.

The next step, in developing an overall methodology is to consider how this phase might relate to the different approaches in Figure 5, in moving from the initial conceptual phase to complete the detailed design.

PHASE 2B - POSSIBLE ROLE OF NUMERICAL CALCULATIONS & LIMIT STATES

The extensive literature on durability and deterioration is mostly concerned with the development of analytical models to predict the processes of deterioration. A key question is: to what extent can these techniques be used, in a design format?

To understand the potential for this, corrosion protection is taken as an example, and a list of possible resistance options produced, which might be considered as part of a multi-layer approach as suggested in Figure 5. A possible list is given in Table 7, under the sub-headings: Materials; Design: Construction. The basic defence strategy would involve a choice of options from each group, and the design skill comes, in assembling the best options in individual cases.-

Most of the Design and Construction options in Table 7, are considered at the conceptual or phase 2a stages of design, in deciding what basic strategy to use, and in making judgements about construction quality. Largely, these are not numerate. However, the materials-based options would also be part of a multi-barrier approach. Here, the designer might wish to evaluate different combinations of, say, concrete cover and quality; in that context, numerical modelling could have a role to play.

Table 7 Shopping list of resistance options for corrosion

A Materials	Concrete quality	-	mix proportions
		-	mix ingredients
	Cover	-	minimum
		-	tolerances
	Permeable formwork		
	Concrete protection	-	sealers
		-	coatings
		-	penetrants
		-	layers
	Rebar protection	-	epoxy coating
	Special steels		
	Non-corrodible reinforcement		
	Cathodic protection		
B Design	Design concept		
	Structural detailing		
	Cladding, services, fittings, finishes		
	Articulation, joints, movement		
	Treatment of water, drainage		
	Control of the environment, barriers		
	Provision for inspection, maintenance, replacement		
	Accurate assessment of effects of deterioration mechanisms		
C Construction	Construction methods		
	Quality control		
	Certification		
	Testing		
	Rationalisation, standardisation, simplification		

The question then is:- should these predictive tools be developed so that they can be used with confidence, in individual designs - or should they be used to develop simplified generic guidance, on the relative merits of alternative options.

If we consider the first of these possibilities, then we would have to satisfy the usual design condition, expressed below in its simplest form:-

$$S_K . g_S \leq \frac{R_K}{g_R}$$

where

S_K is the characteristics value of the aggressive action

γ_S is a partial safety factor for this loading

R_K is the characteristic resistance

γ_R is a partial safety factor for that resistance

To apply this, we need to have:-

1. Detailed knowledge of the loads, pertaining to the particular environment.

2. A model which predicts the penetration mechanism of the aggressive action.

3. Agreed limiting performance criteria, eg. the reinforcement should not be depassivated for x years.

4. Agreed values for the partial safety factors.

Of these four items, serious work has only been done on item 2. Item 1. will always be a major difficulty (see Table 4), except for major structures where detailed investigation can be justified; at the other extreme, detailed analysis is not necessary for indoor concrete.

Where modelling can be justified is in the generic development and evaluation of the relative merits of options in a multi-layer protection system. This is where the research community should now be concentrating its efforts. What sort of guidance are designers looking for here? If we consider the options in Table 7, then, for a given situation, questions such as the following require quantitative answers:

1. How does the use of permeable formwork compare with protective coatings or with adding an extra 20mm of cover? Or

2. Can we add the benefit of a concrete protection system to a Grade 60 concrete, as well as to a Grade 30? Or

3. How long do these additional protective layers last, without having to be renewed - and what is the cost/benefit ratio for each?

It is in this context, in relation to Table 7, that numerical calculations and modelling should first be deployed. Because of uncertainties or variability in environmental loading and quality of construction, designers will frequently be attracted to the multi-layer protection option in Figure 5. If they then look at Table 7, they need some rational basis for making a choice:- how many options to use for different aggressivities; which ones are genuinely compatible and additive; which combination is then the best, in cost/benefit terms?

SUMMARY AND CONCLUDING REMARKS

The design approach outlined earlier will require significant development prior to routine use. What are the essentials, and how might it be introduced progressively in the future? Table 8 presents a summary.

Sustainable construction requires the best use of resources to accurately meet performance requirements for our infrastructure. This is especially true for performance-in-service-with time, where we need to put some effort into defining these requirements better, as well as finding good solutions to satisfy them.

Table 8 Essentials in developing durability design

1	A compatible approach, involving design, materials and construction.
2	Use a specification approach for aggressive actions, which affect the concrete (Table 3).
3	For corrosion, the progression, in time, is (a) Recipes now (Code values for cover and grade). (b) Multi-layer protection in the short-term (Figure 5; Table 7). (c) Complete design framework (longer term).
4	Design must lead in:- (a) Phases 1, 2a and 2b and in Table 6 and Figure 5. (b) Introduce new techniques for dealing with long-term performance, Whole Life Costing, Life Cycle Analysis; Risk Analysis; Value Engineering.

Durability design is fundamental in taking this forward. However, it is argued in this paper that this need not involve numerical calculations comparable with those for structural design and safety. In this context, design does not equate to analysis, and much more emphasis is required on conceptual design (in developing a coherent defensive strategy) and on design details, which both make construction easier (and hence better) and are known to work in terms of providing proper resistance to transport mechanisms.

Prime factors in generating this shift in emphasis are:-

1. Recognition that, for individual cases involving routine structures, it is virtually impossible to obtain reliable and accurate information on the nature and intensity of environmental loads, to justify the apparent precision of existing analytical models. Such models can only be used, within an effective design mode, if matched by comparable precision in the input.

2. The dominant influence of quality of construction, which, if ignored, could invalidate rigorous modelling in durability design.

It is questionable if the durability design solutions outlined here are suitable for inclusion in traditional Codes and Standards, with their heavy dependance on specifications and numerical solutions; design has to be interpreted more generally. What would help future developments considerably are the following:-

3. Codes/specifications which set out performance requirements for different types of structure in some detail, and in sustainability terms.

4. The development work outlined in relation to Table 7, in evaluating, to a common base, the relative merits of alternative resistance options - ideally leading to a register of components and materials with known properties and behaviour, under a range of known environments.

5. The development of a systems approach to construction quality, with defined standards (targets) and clear commitments and responsibilities (as in Table 5).

REFERENCES

1. SOMERVILLE, G. The design life of concrete structures. The Structural Engineer, Volume 64A, N°2. February 1986. pp. 60-71. The Institution of Structural Engineers, London.

2. KROPP, J AND HILSDORF, H K. Performance criteria for concrete durability. RILEM Report 12, 1995. E&FN Spon, London. p327.

3. CEMENTA AB. Durability of concrete in saline environment. Cementa AB, Danderyd, Sweden. May 1996. p 206.

4. BROOMFIELD, J P. Corrosion of steel in concrete: understanding, investigation and repair. E&FN Spon, London, 1997. p240.

5. REINHARDT, H W. (1997). Penetration and permeability of concrete. RILEM Report 16. E&FN Spon, London, 1997. p331.

6. COMITÉ EURO-INTERNATIONAL DU BETON (CEB). New approach to durability design. An example for carbonation induced corrosion. Bulletin 238. CEB, Lausanne. May 1997. p138.

7. SARJA, A. AND VESIKARI, E. Durability design of concrete structures. RILEM Report 14. E&FN Spon, London, 1996. p165.

8. SAKAI, K. Integrated design and environmental issues in concrete technology. E&FN Spon, London, 1996. p308.

9. JAPAN SOCIETY OF CIVIL ENGINEERS (1989). Proposed recommendation on durability design of concrete structures. Concrete Library International No. 14. JSCE, Tokyo, 1989.

10. BRITISH CEMENT ASSOCIATION. Development of an holistic approach to ensure the durability of new concrete construction. Final report to the Department of the Environment on Project 38/13/21 (cc 1031). BCA, Crowthorne. October 1997. p81.

11. SOMERVILLE, G. Engineering design and service life - a framework for the future. Prediction of concrete durability. Proceedings of STATS 21st Anniversary Conference. E&FN Spon, London, 1997. pp58-76.

12. COMITÉ EURO-INTERNATIONAL DE BETON (CEB). CEB-FIP Model Code 1990, Design Code. Thomas Telford, London 1993.

13. BRITISH STANDARDS INSTITUTION. Buildings-Service life planning. Part 1: General principles. Ref. No. HB10141. BSI, London. October 1997. p60.

14. BRITISH CEMENT ASSOCIATION. Durability by Design. Final report to the Department of the Environment (now the DETR) on Project CI.38/13/12 (cc 716). BCA, Crowthorne, UK. April 1998. p46.

15. MUNCH-PETERSEN, C. ET AL. Concrete strategy for the Oresund Crossing. Betonwerk & Fertigteil - Technik. Bft 11. November 1997. pp44-54.

16. DUNASZEGI, L. High performance concrete a key component in the design of the Confederation Bridge. Newsletter, Beton Concrete Canada. Vol 1, No. 3. December 1996. pp1-2.

INDEX OF AUTHORS

SUBJECT INDEX

This index has been compiled from the keywords assigned to the papers, edited and extended as appropriate. The page references are to the first page of the relevant paper.